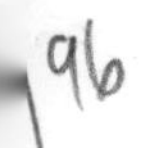

GLOBAL WARMING

The Truth behind the Myth

GLOBAL WARMING

The Truth behind the Myth

MICHAEL L. PARSONS, Ph.D.

Plenum Press • New York and London

Library of Congress Cataloging-in-Publication Data

On file

ISBN 0-306-45083-6

Insight Books is a Division of Plenum Publishing Corporation
233 Spring Street, New York, N.Y. 10013-1578

An Insight Book

10 9 8 7 6 5 4 3 2 1

Printed in the United States of America

To my dad,

O.L. Parsons, M. D. (retired),

for his lifelong love of the environment and support of all pursuits that I have undertaken.

Foreword

The noted political scientist Aaron Wildavsky once termed global warming the "mother of all environmental scares." A phenomenon that may not even be detectable has been linked to all sorts of calamities: collapse of the Antarctic ice sheet, worldwide flooding due to rising ocean levels, disastrous hurricanes, droughts and agricultural disasters, mass starvation, and the spread of tropical diseases putting three billion people at risk. These disasters are not grounded in fact, of course, but spring from the feverish imagination of activists and their ideological desire to impose controls on energy use and to stop—or at least micromanage—economic growth.

With climate change at the top of the international environmental agenda, it is vital that every citizen learn to distinguish reality from hype. This volume fills the need for a balanced presentation of the science and the policy implications of the greenhouse warming issue. It is authoritative, backed by solid data and references, yet presented in a clear and simple manner. The author explains what goes into the forecasts that have captured the media's imagination in the past few years—from the intricacies of climate models to the operation of the sun.

The fearmongers beating the drums have falsely claimed that there is a "scientific consensus" supporting their scenarios: the opposite is more nearly true. The case against panicky policy actions—be they punitive taxes on energy use or more direct control measures—can be encapsulated in three points:

- The model calculations that predict a major warming in the next century have so far not been validated by actual observations of the climate over the past century. More specifically, the precise data from weather satellites since 1979 show essentially zero increase in global average temperature—and, if stretched, at most only one-fifth of the "best" predicted increase.
- Based on solid historical evidence there is every reason to believe that a modest increase in temperature is beneficial for human existence on the planet. Further, most agriculturalists concur that the ongoing increase in atmospheric carbon dioxide will speed up plant growth. What we should fear most is the inevitable return of the next ice age—now overdue.
- Finally, even if greenhouse warming were really here and even if it were as "bad" as hyped, what could realistically be done about it? Most of the world's population lives in developing countries anxious to improve their standard of living, eager to have refrigerators, air conditioners, and cars. Can we deny them their aspirations, while at the same time forbidding the use of nuclear energy, which does not release carbon dioxide into the atmosphere? And what about other greenhouse gases, like methane from rice growing and cattle raising? Stabilization of the *concentration* of atmospheric carbon dioxide requires more than the stabilization of *emissions*: it requires that emissions be cut by 60 to 80 percent—worldwide!

This is not a plea for complacency: any human-induced change in the environment must be carefully monitored and evaluated. In the meantime, common-sense, "no-regrets" policies, like energy conservation, rather than hasty actions based on insufficient science, are in order.

S. Fred Singer, Ph.D.
Director, Science & Environmental Policy Project
Fairfax, Virginia

Preface

Global warming has received much attention in the past few years in the news, popular magazines and books, the classroom, and on television information shows. Predictions from computer climate models have indicated the possibility of rapid warming of the earth and of serious consequences, such as rising sea levels, loss of biodiversity, and destruction of agriculture. Many exaggerations and false statements have been made about the extent of the warming, its causes, and its consequences, which have led to misunderstanding about the nature and extent of the problem. An entire set of beliefs has grown up around both the facts and the misconceptions and has become accepted as truth among the general public. This popular view has been perpetuated by the media, the educational system, the environmental movement, government, and some climate scientists.

We like to use the term "myth" to describe this view; Webster's Collegiate Dictionary defines myth as "a belief or set of beliefs, often unproven or false, that has accrued around a person, phenomenon, or institution."

Myths and mythology are important to society, helping us deal with the world and its complexities. Myths have developed in most cultures throughout history, and they have been based on both observation and speculation, on both experience and superstition. Many myths sought to explain natural forces in terms of gods and goddesses. The ancient people of Hawaii explained the formation of their islands in the story of Pele, a petulant goddess who caused earthquakes and volcanic eruptions. The ancient Greeks attributed earthquakes to Poseidon, the god of the sea. And every culture has had a story of creation that sought to explain the origins of the earth and determine the role of humans in nature.

What is the basis for the myth of global warming? Is global warming a peril to humanity or is the myth a volatile overreaction to scientific speculation? Do the scientific facts support the position that greenhouse gas emissions caused by human activities constitute a problem that warrants global political action? While there is no question that there are greenhouse gases, and there is no question that the greenhouse phenomenon exists, there are many uncertainties about the assumptions used in the computer models that predict global warming. In addition, there are other natural forces at work on our planet that are probably far more important than greenhouse gases, often having an opposite effect on the climate.

This book will present a discussion of global climate and the greenhouse effect, the computer models that are used to predict global warming, the source and balance of the most important greenhouse gases, and the factors left out of the greenhouse models. It will discuss the hysteria generated by an overreaction to scientific speculation and the resulting governmental policy implications. This is not an anti-environment book. Rather, it will attempt to put the facts associated with global climate into proper perspective with other environmental problems, such as overpopulation, depletion of nonrenewable energy resources, and pollution. It will present the scientific facts on both sides of the various issues and document them so that you will be able to form an intelligent and unbiased opinion.

Michael Parsons

San Dimas, California

Acknowledgments

It has always bothered me when people jump on a bandwagon without knowing where it is going. This seems to be the case with most environmental books; in particular, many authors choose to present evidence for only one side of the issue. In April of 1993, when my wife Ginger and I vacationed in Greece, for recreational reading I took along *Our Angry Earth* by Isaac Asimov and Frederik Pohl. I admire Asimov, but this book was one-sided, full of misstatements, exaggerations, and misconceptions, and I was outraged! I started the outline for *Global Warming: The Truth behind the Myth* on that trip.

My wife deserves to be acknowledged as a coauthor of this book and not hidden here in the acknowledgments. She helped research many areas of the book. She ruthlessly edited the entire manuscript, and wrote or rewrote significant portions. In addition, she kept me from writing a "technical" book, which is a tendency of mine because my writing career up until now has been largely scientific. She must have been happier than anyone that "February is finally here!" (February 1995 was the deadline for the manuscript.)

I am fortunate to live in Southern California with its wealth of libraries. I have made use of no less than thirteen of them. The patient librarians had no idea of my project, and I want to thank them for being there.

I also traveled down the "information superhighway" in preparing this book, using both DIALOG and CompuServe on-line computer services for finding articles that were not otherwise accessible. I even had a conversation with Tom, the "SOS" rescue computer-person at CompuServe—Thanks for the help, Tom!

Government agencies were also helpful. The DOE's Carbon Dioxide Information Analysis Center (CDIAC) at Oak Ridge National Laboratory supplied me with invaluable reports in a very timely manner. I particularly want to thank Ms. Laura Morris of CDIAC, who sent and even faxed important information when I requested it. NOAA's National Geophysical Data Center also supplied important data, and the U.S. Department of Agriculture and NASA were helpful as well.

I would like to thank several members of my family who read and made helpful comments on portions of the manuscript, including Guinn Sherlock, Erin and Fred Fugere, and Bill Thomas.

Finally, I thank Insight Books for taking on this project and editor Frank Darmstadt for his encouragement and guidance.

I hope that I have fairly presented the facts that are known on the issue of global warming and, most important of all, I hope that you learn and enjoy reading about this fascinating issue.

Contents

PART III. FACTORS AFFECTING CLIMATE AND CLIMATE COMPUTER MODELS

PART IV. PREDICTION OR REALITY?

Acronyms and Abbreviations

AAAS	American Association for the Advancement of Science
ACRIM I	Active Cavity Radiometer Irradiance Monitor
CCN	cloud/condensation nuclei
CFC	chlorofluorocarbon
CH_4	methane
CO_2	carbon dioxide
COADS	comprehensive ocean–atmosphere data set
DDT	dichlorodiphenyltrichloroethane
DMS	dimethyl sulfide
DOE	Department of Energy
EEC	European Economic Community
ENSO	El Niño/Southern Oscillation
EPA	Environmental Protection Agency
GIGO	garbage in–garbage out (refers to computer data input–output)
GCM	general circulation model
GNP	Gross National Product
GWP	global warming potential
HCl	hydrogen chloride, or hydrochloric acid
ICSU	International Council of Scientific Unions
IPCC	Intergovernmental Panel on Climate Change
IR	infrared radiation

MIT	Massachusetts Institute of Technology
MSU	microwave sounding unit
NAS	National Academy of Sciences
NASA	National Aeronautics and Space Administration
NCAR	National Center for Atmospheric Research
NOAA	National Oceanic and Atmospheric Administration
NSF	National Science Foundation
NWS	National Weather Service
OMB	Office of Management and Budget
OPEC	Organization of Petroleum Exporting Countries
OSO	Orbiting Solar Observatory
OTA	Office of Technology Assessment
PPM	part per million
SO_2	sulfur dioxide
UN	United Nations
UNCED	United Nations Conference on Environment and Development
UNEP	United Nations Environment Programme
UV	ultraviolet radiation
WMO	World Meteorological Organization

PART I

The Myth of Global Warming

ONE

The Greenhouse Effect, Global Warming, and the Role of Computers

There was Chaos at first, and Darkness and Night. In the vast hollows of Darkness, gold-winged Love burst forth like a whirlwind. Love mingled in Darkness and Chaos, then Love blended the universe and brought forth the fair Earth, and the Sky, and the limitless Sea, and the race of the Gods, who shall never die.

—GREEK CREATION MYTH ADAPTED FROM ARISTOPHANES' *THE BIRDS*

In the beginning God created the heaven and the earth. And the earth was without form, and void; and darkness was upon the face of the deep. And God said, Let there be light: and there was light. And God said, let the waters under the heaven be gathered together unto one place, and let the dry land appear. And let us make man in our image. And God saw everything that he had made, and, behold, it was very good.

—JUDAEO-CHRISTIAN CREATION MYTH ADAPTED FROM THE *BIBLE, KING JAMES VERSION*

When life first appeared on our planet, whether by chance or design, it was possible only because the climate was friendly; it was neither too hot nor too cold. The atmosphere, containing water vapor, carbon dioxide, and other components, acted like the glass panels of a greenhouse, letting the sun's heat in but preventing some of it from escaping, thus warming the planet. If it were not for this greenhouse effect, temperatures at the earth's surface would be far colder than they are, and life as we know it could not exist.

Global warming has become an issue of concern because of the perception that increasing greenhouse gases will cause the earth to warm so fast that nature (and humankind) may not be able to adapt to the rapid change. The atmosphere has always acted as a warming blanket around the earth. It is not the warmth of this blanket but the possibility of an abrupt or extreme temperature change that is causing some scientists concern.

The popular view of global warming—the "myth"—has developed over the past decade because of predictions of computer climate models. This view has been presented in books by authors such as Al Gore, Stephen Schneider, Isaac Asimov, and Paul Ehrlich, and it has been advanced by the media and in school classrooms. Environmental organizations have taken up the cause and demanded political action. At times the reaction to this view reaches a feverish pitch, generating hysteria. Chapter 2 covers this phenomenon and examines some reasons for it.

What is the global warming myth? It is a scenario of doom that contains certain facts and also certain assumptions:

- Atmospheric carbon dioxide (CO_2) causes warming of the planet.
- Man's activities are increasing the amount of carbon dioxide.
- The average temperature of the earth has increased approximately 0.5°C (0.9°F) in the last 100 years.
- Global temperature will increase another 1.5–4.5°C (2.7–8.1°F) by sometime in the next century if we do not take drastic measures.

- The predicted results of this warming include melting of the polar ice caps, flooding of coastal cities, massive extinction of species, and the deterioration of civilization as we know it.

What is the basis for the myth? Which components of the myth are proven facts, and which are assumptions based on computer modeling?

GLOBAL WARMING: THE BACKGROUND

The issue of global warming first emerged when scientists became aware of the amount of carbon dioxide (CO_2) being added to the atmosphere as a result of human activity. However, the first scientists to recognize the relationship between carbon dioxide and climate were concerned about excessive cooling, not warming.

Jean-Baptiste Joseph Fourier, a French scientist during the Napoleonic years, is generally given credit for being the first to describe the greenhouse effect in the 1820s. Fourier was known for both his developments in mathematics and his studies of Egypt. He studied the properties of heat—its radiation and transfer—and he developed the mathematics of partial differential equations. His discovery of the greenhouse effect was a result of his heat studies.

John Tyndall, another scientist interested in the greenhouse gas phenomenon, was an Irish physicist and Professor of Natural Philosophy at the Royal Institution in London. (Tyndall is best known for his studies of light scattering from the earth's atmosphere, which led to our understanding of why the sky is blue.) In the 1860s, Tyndall measured the radiation absorption efficiencies of various gases, a measure of their effectiveness as greenhouse gases. He was concerned that a decrease in atmospheric CO_2 could lead to another ice age.

In 1896 Svante August Arrhenius theorized that carbon dioxide was a greenhouse gas being introduced into the atmosphere by the burning of carbon-based fossil fuels. Unlike some of today's scientists, he concluded that any warming caused by this effect was good for the human race. He looked forward to a warmer climate that would bring more abundant crops "for the benefit of rapidly propagating mankind."[1] At the time, no one got particularly excited about his global warming theories. Arrhenius, who taught himself to read at the age of three, did not always operate within the popular framework of scientific research, being accused of, among other things, manipulating imaginary data. Ultimately, many of his ideas began to be accepted. In 1903 he became the first Swede to win the Nobel prize, and he is now considered the founder of physical chemistry. Even so, some of his theories remain outside the realm of acceptance, such as his suggestion that life is spread by bacteria activated by collisions of stars.

Little was written about carbon dioxide as a greenhouse gas for the next forty years. In 1938 G. S. Callendar, an English meteorologist, evaluated historical records of atmospheric carbon dioxide and showed a trend of increasing concentration.[2] He, too, welcomed the potential future warmth and felt that the carbon dioxide would provide fertilizer for increased farm production. He hoped that the return of the deadly glaciers would be delayed indefinitely.

Eventually, climatologists became concerned that the increase in carbon dioxide would cause too much warming. In 1956, Gilbert Plass, a scientist at Johns Hopkins University, Baltimore, Maryland, expanded on a theory that attributed great importance to carbon dioxide. He suggested that atmospheric carbon dioxide controls the climate, and he predicted that accumulated atmospheric carbon dioxide from mankind's burning of fossil fuels would cause a global temperature increase of 1.1°C (2.0°F) by the beginning of the twenty-first century.[3] It was well known by then that large quantities of carbon dioxide were being emitted into the atmosphere because of the fuel used to meet the energy and industrial requirements of society.

Increases in atmospheric carbon dioxide were actually observed about this time by various scientists working independently of each other. The most prominent was Charles David Keeling of Scripps Institution of Oceanography, La Jolla, California. Keeling measured the atmospheric carbon dioxide concentration at various locations in the United States and found that it had increased from about 275 parts per million (ppm) in the nineteenth century to 311 ppm in 1956.[4] [A part per million of CO_2 means one molecule of carbon dioxide for every million (1,000,000) molecules of total atmospheric gases.]

In 1967 the warnings became dire (and much more publicized) as Syukuro Manabe and Richard Wetherald, climatologists working at Princeton University, predicted increases of about 2°C (3.6°F) by the end of the twenty-first century.[5] This prediction was based on one of the first of many modern complex computer models. Ironically, this study was reported during a time when many scientists, observing a global cooling trend from about 1950 until 1980, were predicting the beginning of another ice age.

During the 1980s, computer modeling of the earth's climate and the effect of doubling atmospheric carbon dioxide became increasingly popular among climatologists. This modeling activity and heavy speculation concerning the consequences of the predicted warming resulted in enhanced coverage by the scientific press, including the weekly general science journals such as *Science* and *Nature*. The popular press had a heyday with these speculations (Chapter 2).

Publicity about the imminent danger of global warming was brought to a head in 1988, one of the hottest years on record, when James Hansen, director of NASA's Goddard Institute of Space Studies in New York City, testified before Congress that global warming had begun! In 1989 he testified again and defended his testimony in an interview with Richard Kerr, a research newswriter for *Science*, stating:

> I said three things. The first was that I believed the earth was getting warmer and I could say that with 99% confidence. The second was that with a high degree of confidence we could associate the warming and the greenhouse effect. The third was that in our climate model, by the late 1980s and early 1990s, there's already a noticeable increase in the frequency of drought.[6]

His statements were so strong that the Office of Management and Budget required that they be qualified with a statement about the uncertainties of computer models.

At the time of Hansen's 1989 congressional testimony, the Workshop on Greenhouse-Gas Induced Climatic Change was being held in Amherst, Massachusetts. Most of the climatologists at the workshop felt that Hansen had gone too far, overstepping the boundaries of good science with his pronouncements. The forty participants who were present on the last day of the workshop issued a press release that stated:

> It is tempting to attribute [the 0.5°C warming of the past 100 years] to the increase in greenhouse gases. Because of the natural variation of temperature, however, such an attribution cannot now be made with any degree of confidence.[7]

Scientists rarely like to state uncertain events in absolute terms, and it is not surprising that this group of peers was upset with Hansen for publicly stating speculation as fact.

Even Stephen Schneider, a prominent climate modeler at the National Center for Atmospheric Research and proponent of the global warming theory, pointed out the weakness of Hansen's position when he stated, "He's not running a realistic ocean." This is a reference to the computer model of the ocean that is incorporated into the climate model. Schneider went on to say, "They [Hansen's group] have been using a pretty hokey ocean, we all have. But you have to have less confidence because of that."[8] In spite of these statements, Schneider still believes that global warming is beginning and will cause major problems in the future.

In Chapter 4 we focus on the computer climate models and explain how they work, discuss the uncertainties associated with their input data, and evaluate their predictions. There we will discuss the "hokey ocean" and other simplifications that are common practice in computer modeling.

There is little controversy over the greenhouse effect as a scientific theory. The controversy arises over the global warming predictions: the causes (whether or not they're human induced), the amount of warming and the timing of it, the implications for life and society, and most of all, what to do about it.

Let us take a more detailed look at the greenhouse effect, global warming, and scientific prediction.

GREENHOUSE EFFECT

The earth derives most of its energy from the sun, which continuously emits radiant energy equivalent to a 6,000°C (about 11,000°F) boiling caldron. This radiant energy, a very small portion of which reaches our stratosphere, consists of the full range of radiation, including the ultraviolet (UV), visible, and infrared (IR) portions of the sun's spectrum (Figure 1.1).

On the one hand, ultraviolet radiation, the most energetic portion, causes damage to living tissue if it is overexposed; it is UV radiation that causes sunburn and blinds people who stare directly at the sun. On the other hand, the interaction of UV and visible radiation within plants provides the energy required for the photosynthesis reactions essential to plant growth. Since plants are the foundation of our food chain, this radiation is essential for all life.

The visible portion of the sun's radiation is simply that to which our eyes are sensitive. We cannot see the UV or IR portions of the sun's spectrum, but we can feel the warmth of IR radiation—it is the same as that which is radiated from an infrared heat lamp. The IR energy is too low to trigger our optical nerve endings, so our eyes are not sensitive to this radiation.

The sun furnishes the earth with a generous amount of ultraviolet, visible, and infrared radiation because it radiates at such a high temperature. The earth in turn radiates some of the sun's energy back into space at a very low temperature, about 15°C (59°F).[9] Because of this low temperature, nearly all of the earth's radiation is in the low-energy infrared region.

If there is a balance between the energy provided by the sun and that returned to space by the earth, the earth's temperature will remain constant. If the earth retains more energy than it returns to

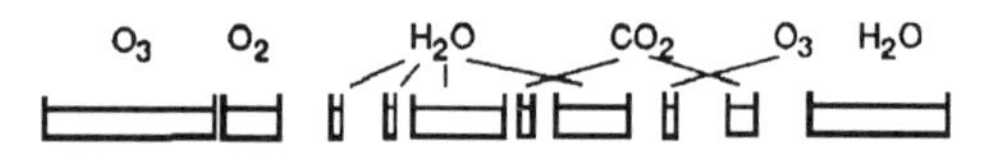
Atmospheric absorption bands
O3
O2
H2O
CO2
O3
H2O

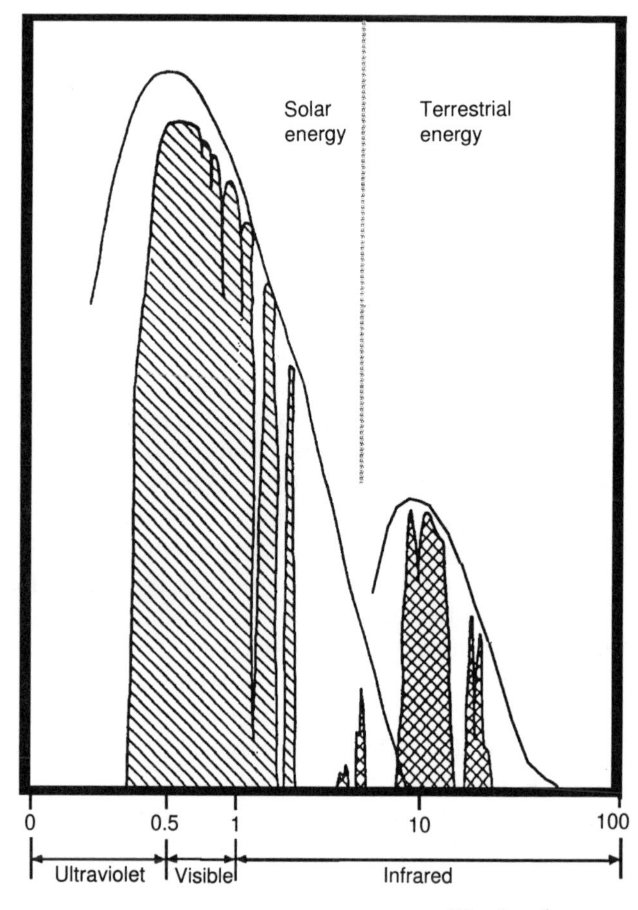
Energy flux density
Solar energy
Terrestrial energy
0
0.5
1
10
100
Ultraviolet
Visible
Infrared
Wavelength, µm
Radiation from a body at 6000 K (e.g., the Sun)
Direct solar radiation at the Earth's surface
Infrared emission from the Earth

space, it will heat up; conversely, if the earth retains less energy than is returned to space, it will cool down.

When the sun's radiation enters the atmosphere, many things can happen to it. When sunlight encounters dust or clouds or any other type of matter, it may be absorbed, reflected, scattered, or transmitted. Some of the radiation passes through to the earth; some of it is reflected back into space; some of it interacts with oceans or plants and is absorbed; some is scattered or reflected from the ice and snow cover.

While solar radiation can interact with all matter, how it acts depends on the portion of the solar spectrum to which it belongs: UV radiation has higher energy than IR radiation, and visible radiation is in between. A good example of this interaction occurs in the ozone layer in our stratosphere. The ozone layer is very efficient at absorbing very high-energy UV radiation, but it lets much of the lower-energy UV and most of the visible and IR radiation pass right through into the lower atmosphere, or troposphere (see Figure 1.2). This radiation ultimately interacts with clouds or dust in the troposphere or with the earth itself, causing the earth to heat up proportionally to the energy that is transferred.

FIGURE 1.1. Characteristics of solar radiation reaching the earth and infrared radiation emitted by the earth. This diagram shows the energy intensity of solar radiation reaching earth's atmosphere (the solid curve on the left) and the energy intensity of the earth's radiation (the solid curve on the right). The shaded areas indicate the radiation that is *not* absorbed by the molecules in the atmosphere. The most important active molecules that take part in the greenhouse effect are indicated above the curves. These are the molecules that absorb the radiation both entering the earth's atmosphere and being emitted by the earth itself. These molecules are ozone (O_3), water vapor (H_2O), and carbon dioxide (CO_2). Note that the sun's radiation includes the ultraviolet, visible, and infrared regions of the spectrum, whereas the earth's radiation includes only the low-energy infrared region. This figure was adapted from E. G. Nisbet, *Leaving Eden: To Protect and Manage the Earth* (Cambridge: Cambridge University Press, 1991), p. 14. (Reprinted with the permission of Cambridge University Press.)

As stated above, the warmed earth also radiates energy, but at a very low temperature as compared with the sun's temperature. The earth's temperature currently averages about 15°C (59°F), whereas the sun's surface is 6,000°C (11,000°F). The earth's radiation is in the low-energy IR range and can be absorbed by certain components of the atmosphere, such as clouds, particulate dust, or greenhouse gases. We will look at these issues in more depth in later chapters.

It is important to understand the nature of the atmosphere that blankets the earth. The classification of the different layers of the atmosphere along with the temperature variation related to each layer is shown in Figure 1.2. As altitude increases, pressure decreases exponentially. There is insufficient oxygen to support life at the top of the troposphere, which is about the altitude where commercial aircraft fly. It is estimated that 99% of the mass of the atmosphere is in the first 30 kilometers (about 20 miles) of the ground. The mean diameter of the earth is 12,742 kilometers (almost 8,000 miles), so the atmosphere is quite thin compared to the size of the earth. The focus of the global warming issue is on the troposphere and, to a limited degree, the stratosphere.

FIGURE 1.2. Classification of atmosphere and idealized variation of temperature with altitude in the atmosphere. The troposphere is the atmosphere closest to earth. Note that its temperature steadily decreases with altitude to –60°C (–76°F), which will freeze any moisture in the atmosphere. Therefore, very little water vapor escapes to the stratosphere. The major greenhouse effect occurs in the troposphere. The stratosphere contains the ozone layer, which is the focus of the "ozone hole" over the Antarctic. Volcanic activity contributes to the sulfuric acid aerosol in the stratosphere (Chapter 9). Ninety-nine percent of the mass of the atmosphere is below 30 kilometers (about 100,000 feet). The mesosphere and thermosphere have very low atmospheric pressures and are essentially outer space for the purposes of climate discussions. The tropopause, stratopause, and mesopause refer to the conceptual separations of the various atmospheric layers. This figure was adapted from the data in Jose P. Peixoto and Abraham H. Oort, *Physics of Climate* (New York: American Institute of Physics, 1992), pp. 14–15.

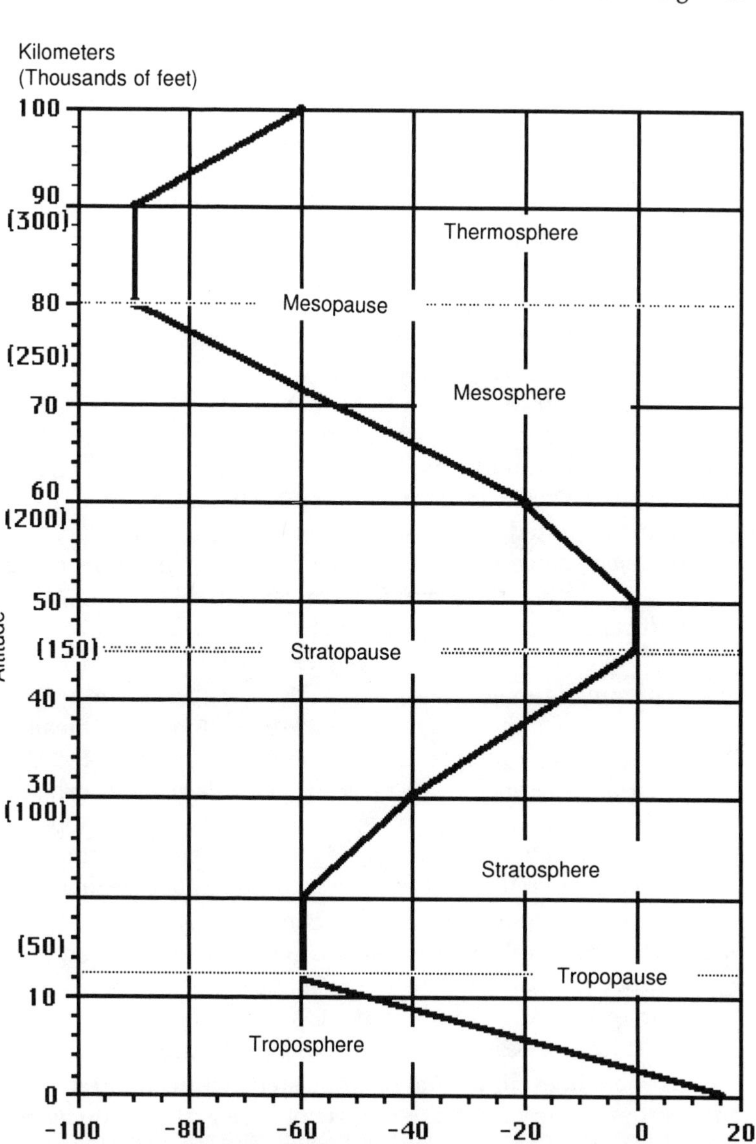
Kilometers
(Thousands of feet)
100
90
(300)
80
(250)
70
60
(200)
50
(150)
40
30
(100)
(50)
10
0
Altitude
Thermosphere
Mesopause
Mesosphere
Stratopause
Stratosphere
Tropopause
Troposphere
-100
-80
-60
-40
-20
0
20
Temperature (°C)

GREENHOUSE GASES

The components of the earth's atmosphere that trap radiation are called "greenhouse gases." The major components of our troposphere—oxygen, nitrogen, and argon—are transparent to radiation, and thus they are not greenhouse gases. Only the minor and trace components are capable of interacting with solar radiation.

The most important greenhouse gas is water vapor; others include carbon dioxide, hydrocarbons (such as methane gas), various oxides of nitrogen and sulfur, and many man-made chemicals (such as chlorofluorocarbons, or CFCs).

Most of the greenhouse gases have natural sources; they come from volcanic eruptions, ocean evaporation, and animal and plant respiration. For example, the "smoke" often seen over the forests of the Smoky Mountains is mostly composed of natural organic compounds emitted by trees. But there are also man-made, or *anthropogenic*, sources. When any fuel is burned, for necessity or for convenience, carbon dioxide and water vapor form as combustion by-products. The fuel can be wood, petroleum, natural gas, coal, or animal dung, but the by-products are essentially the same.

People are CO_2-manufacturing machines. We breathe in oxygen from the air, convert it into carbon dioxide in our lungs, and breathe the carbon dioxide out. (Plants do just the opposite—they breathe in carbon dioxide and breathe out oxygen.) Each adult exhales about 1.5 liters of carbon dioxide per minute. The earth's population is over 5.2 billion, so there is a direct and significant input of carbon dioxide into the atmosphere just from human breathing!

Automobiles emit significant amounts of carbon dioxide, as well as trace amounts of carbon monoxide, oxides of nitrogen, and hydrocarbon gases. Electric power generation by means of coal, oil, or natural gas produces carbon dioxide and sulfur oxides. Many manufacturing processes create gaseous by-products as well.

It is sometimes difficult to distinguish between "natural" and "man-caused" carbon dioxide emissions. For example, domestic animals are bred for work, for food, or as pets; they, like humans, breathe carbon dioxide into the atmosphere. Are domestic animals, therefore, anthropogenic sources of carbon dioxide? Which forests are natural and which ones are grown by humans for wood?

Water vapor is also generated by all of these sources; however, the water that evaporates from oceans and lakes far exceeds that produced by any anthropogenic source. In fact, water vapor accounts for *most* of the greenhouse effect. According to the Intergovernmental Panel on Climate Change (IPCC), a large group of environmental scientists set up by the World Meteorological Organization and the United Nations Environment Programme to study climate change, water vapor is responsible for well over 60% of the greenhouse warming in our atmosphere.[10] The hydrological cycle and the complexities of water in our environment are so important in the global warming discussion that they warrant a complete chapter (Chapter 8).

Since the mid-fifties, environmental scientists have been investigating the increase in greenhouse gases, but they have not focused on the natural levels or sources of greenhouse gases. They are mainly interested in the greenhouse gases introduced into the atmosphere by man's activities—the anthropogenic effect. While anthropogenic causes are the only ones that we can control to any great extent, it is important to remember that they constitute a very small part of the total effect. Carbon dioxide is not the only gas under scrutiny. Methane, nitrous oxide, ozone, CFCs, and others are considered greenhouse gases as well; these are all discussed in Chapter 6.

GLOBAL TEMPERATURE INCREASE

The IPCC has stated that, because of the anthropogenic increases in greenhouse gases during the past century, the average global temperature of the earth has increased by 0.3–0.6°C or about 0.5°C (0.9°F).[11] But consider this: The earth's temperature often varies by 30–40°F from day to night, and the seasonal variation changes the highs and lows an additional 20–40°F. How then can we detect so small a temperature change over a hundred-year period? The answer is: We probably cannot.

Another question about global temperature is: Has the temperature in fact increased? Methods of temperature measurement have changed during the past 100 years. Instruments have improved and procedures have changed. In many instances in the past, instruments were calibrated to different standards. At times, measurements were taken at different locations. Some scientists believe that measurements from satellites are the only accurate ones. Satellite measurements have been made for a short time, only fifteen years, and so far have shown no increasing temperature trends. An entire chapter of this book is devoted to the problems of measuring global temperature (Chapter 5). Is it possible that scientists are kidding themselves about this 0.9°F temperature change altogether?

Despite the inability to detect a global temperature change of less than 1°F during the past 100 years, we do know that the earth has experienced both colder and warmer climates during its long history. There were several ice ages during the period of time scientists are able to study by means of tree ring and ice core research. Every hundred thousand years the global temperature has decreased and great glaciers have covered much of the Northern Hemisphere. During the past few centuries there was a cooling effect that was significant enough for scientists to name it the Little Ice Age. It is possible that the presumed 0.9°F warming during the past century is the earth's natural recovery from the Little Ice Age.

VARIATION IN SOLAR ENERGY AND THE EARTH'S ORBIT

The radiation flux from the sun to the earth is considered a constant, or nonfluctuating, value in virtually all computer climate models. However, studies of the sun have shown an eleven-year cycle in the level of the sun's energy output. There is increasing evidence of longer cycles as well, as we discuss in Chapter 7. Since most of the earth's energy is directly derived from the sun's energy output, we must consider that part of the perceived 0.9°F increase during the past century might be due to variation of the sun's energy output.

In addition, the earth's orbit (distance from and inclination toward the sun) varies with time, changing the amount of radiational energy reaching the earth. Due to the change in the earth's orbit from summer to winter, the radiation reaching the earth varies by 7%, a variable that is included in most models. However, there are additional orbital variations that have been shown to correlate with the ice ages of the past. These variations are further discussed in Chapter 5 with respect to the earth's past temperature variations.

GLOBAL WARMING MODELS

Thanks to the incredible computational power now available with supercomputers, scientists can build mathematical models that simulate all types of processes, such as the earth's air circulation and the ocean currents and circulation. A concentrated effort has centered on weather forecasting and on determining the average global temperature and its variations. Many factors must be included: the effects of the earth's rotation, the moon's gravitational pull, wind, clouds, ocean currents, ice coverage, variation of the sun's radiation output, variation in the earth's orbit, greenhouse gases (both natural and anthropogenic), day–night temperature fluctuations, and seasonal temperature fluctuations, as well as major natural events such as volcanic eruptions, major forest or grass fires, hurricanes, etc. One must truly sit in awe at the ambition of such a project.

Global warming models have been developed and refined by environmental scientists for many years. Until very recently many of the input factors listed above were oversimplified or ignored in order to decrease expensive computer time and because the scientists simply did not know how to include them in their models. Many different scientific groups have worked on the problem, and consequently, many different approaches have evolved.

Not surprisingly, various models predict different results. The IPCC's 1992 report provides results from seven different computer

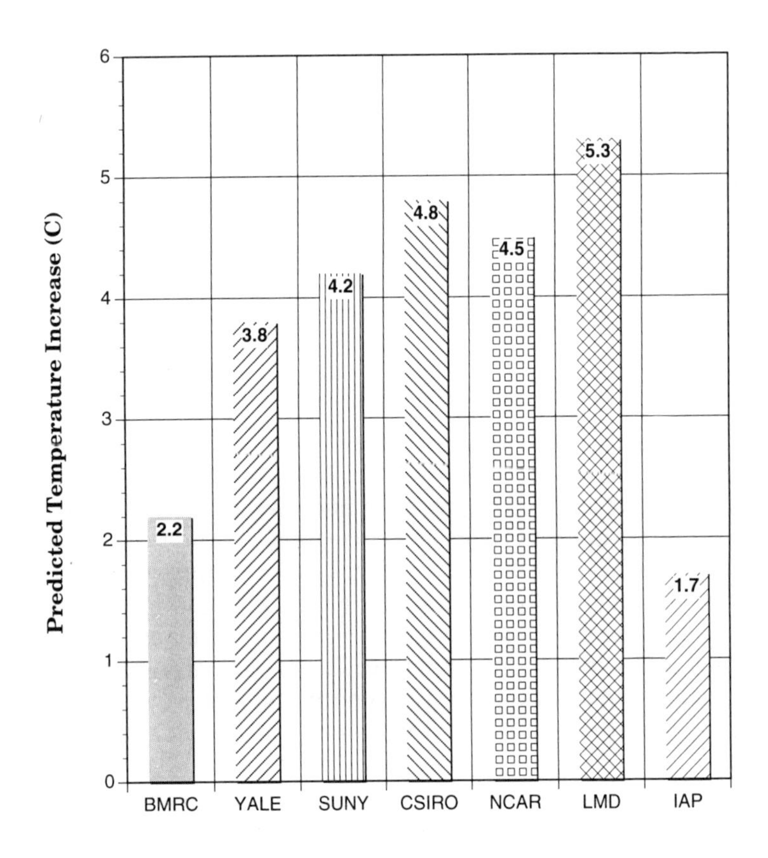

FIGURE 1.3. Predicted global warming (°C) based on doubled carbon dioxide levels determined by seven different general circulation models (GCMs). The symbols BMRC, YALE, etc. simply identify the different models. Adapted from data in J. T. Houghton, B. A. Callander, and S. K. Varney, eds., *Climate Change 1992* (Cambridge: Cambridge University Press, 1992), p. 111.

climate models. These models, which were more sophisticated than the ones available for the 1990 assessment, were used to calculate global temperature, assuming a doubling of carbon dioxide in the atmosphere. (This means that the climate modelers first used current carbon dioxide levels, and then they ran the same models doubling the carbon dioxide values.) The predicted levels of global warming varied by over 300%! These results are illustrated in Figure 1.3. Temperature increases were calculated by the different models to be anywhere from 1.7–5.3°C (3.25–9.54°F).[12] Many scientists believe that such large variation in the results of different models renders the predictions from any of them virtually useless. We take a more thorough look at the factors affecting global warming and the simplifications to the global warming models in later chapters.

SCIENTIFIC PREDICTION

When a scientist formulates a theory, that theory should explain how something works now or how it will work in the future. For example, early in the seventeenth century, when Galileo confirmed Copernicus's theory that the sun, not the earth, was the center of our solar system, he was able to explain the observed movement of the planets through his use of the newly invented telescope. Galileo was scorned by the Church at the time and even forced to recant this theory, but the Church later accepted his observations as truth in the face of overwhelming evidence. By contrast, many scientific theories and subsequent predictions have been proven to be either wrong or inadequate, based on additional, more accurate observations. This was the case with Ptolemy's earth-centered theory of the universe, the theory Copernicus and Galileo overturned.

The myths of many ancient civilizations associated solar and lunar eclipses with death or the devouring of a heavenly body by some animal or monster. In some cultures the priests or holy men were also astronomers; part of their success centered on their ability to predict the feared eclipse events. Historical records from cultures of most parts of the earth indicate the importance of eclipse prediction. One of the few surviving written documents from the Mayan civilization, the Dresden Codex, predicted all of the seventy-seven partial or total solar eclipses in the period A.D. 755–788.[13] Predicting the disappearance of the sun must have seemed magical to the Mayans, who were just beginning to understand the technology of farming. They knew that the sun was essential to their survival, and when the priests demanded it, they even made human sacrifices to appease this mighty "Sun God."

Scientists have developed increasingly sophisticated prediction capabilities using mathematical models to express scientific theories. An excellent example of these models is the aircraft flight simulator, which predicts an aircraft's reaction to the pilot's commands. Pilots can obtain hours of "flight experience" without ever leaving the ground.

One of the most important accomplishments of this century was the development of the atomic bomb and the subsequent peaceful use of nuclear energy. This development was the direct result of successful mathematical modeling of the atomic nucleus and the ability to predict the energy produced by nuclear reactions. The "critical mass" of uranium required to sustain a nuclear reaction and the geometry and design of the implosion required to achieve this critical mass both were made possible by sophisticated mathematical models. Hans Bethe, one of the scientists responsible for developing the atomic bomb, first recognized that these nuclear reactions are the same as the reactions taking place in the sun, which ultimately provide the earth's energy. As a consequence, we now have a more accurate model of the sun.

Another result of successful computer modeling is the placing of satellites in space the by United States National Aeronautics and Space Administration (NASA). NASA scientists must calculate the speed and position required to place a satellite in a specific orbit. The force of gravity from the earth, the sun, and the moon vary tremendously in their effects on the orbit of a small satellite, and these forces must be accounted for in a model of the satellite's orbit. In some cases the goal is for the orbit to be around the equator, in others from pole to pole, and finally, for certain applications, it is desirable that the satellite remain in one spot and not move with respect to the earth. Different models are required for each type of orbit. The scientists at NASA have launched many missions to the sun's planets, the earth's moon, and around the earth, and they have placed satellites with great precision because of their ability to model all these factors accurately.

The need for ever greater mathematical capabilities has driven the development of computer technology to the point where it is now possible to make millions of mathematical manipulations per second. This computational ability allows the development of models that reach far beyond our rational grasp. In a matter of minutes it is possible to solve hundreds, even thousands, of simultaneous equations that would take literally years to solve by hand. Scientists have grown dependent on the computer for the answers to many scientific questions.

APPLYING MODELS: THE LIMITATIONS OF SCIENTIFIC PREDICTION

Computer models are very important to scientists for the simulation of many processes. A major purpose of this book is to show the limitations of computer models and thus enable the reader to judge the validity of the predictions of the global warming models. These models are so complex that they stress the current state-of-the-art computers' capabilities, and yet there are still many aspects of the models that are woefully inadequate at simulating the realities of our global system. (The computer models and their capabilities and limitations are discussed in detail in Chapter 4.)

There are several rules that computer modelers should not violate:

1. The data on which the model is built must be realistic.
2. All factors that affect the results must be included.
3. The results predicted by the model must be validated by successful prediction of events within the input data base.

4. The predictions of the model should not be extrapolated very far beyond the limits of the input data.
5. Arbitrary "adjustment" factors must never be used.

This book will show that many of these rules have been broken in computer climate modeling. Here is a prime example:

In 1993 Syukuro Manabe and Ronald Stouffer of Princeton University published a report of their computer climate model, extrapolating the results 500 years into the future.[14] (You may recall that it was Manabe who predicted carbon dioxide increases in 1967.) Making predictions for even 100 years in the future is tantamount to gazing into a crystal ball! Manabe and Stouffer have not validated their model adequately to trust projections of 100 years, much less 500! (Read the example of the marathon runner in Chapter 4 and note the dangers of extrapolating too far beyond the data used to develop a computer model, and you will see the folly of such extrapolations.) Yet science writers of several technical journals, including *Science* and *Nature*, commented on the results as if they were actual experiments.[15] The tone of the articles is reflected in the *Science* article title: "Tales of the Coming Mega-Greenhouse."

THE FUTURE OF SCIENTIFIC PREDICTION

Predicting cataclysmic natural events such as volcanic eruptions, earthquakes, hurricanes, and tornadoes would be beneficial to society, and scientists have made progress in using computer modeling for these purposes. Great strides have been made in weather forecasting. In fact, meteorologists can generally make accurate predictions of the weather at a particular location twenty-four hours or so in advance. However, the twenty-four-hour predictions cannot always be relied on. For instance, in the wake of severe rainstorms and flooding in Southern California in January 1995, weather forecasters published a prediction on January nineteenth for a 100% chance of rain by noon the next day. As the clouds moved in slowly the next morning, the forecasters changed their predictions from 100% to 80% to 60% to 30%, and little or no rain fell until that evening. When meteorologists attempt five-day forecasts, they have very little chance of being accurate beyond the first day or two; the fifth-day predictions are rarely correct.

Meteorologists can pinpoint the landfall of a hurricane within a hundred miles or so several hours before it hits land, making it possible for people to protect their property and flee if necessary.

Earthquakes and volcanoes are the subject of a great deal of research, but scientists are a long way from predicting them with any degree of probability. At best, lessons of preparedness can be learned.

BALANCING POLICY WITH KNOWLEDGE

You will realize after reading this book that no one is certain that human activities are the cause of the global warming perceived to have taken place during the past 100 years. Indeed, you should wonder if in fact the earth has warmed at all during this time and if the ability to measure such a minute change is possible.

Some scientists and many environmental activists are asking, no, they are demanding, that the governments of the world take drastic action to stop all activities that might be contributing to global warming. Major national and international action plans and treaties have been formulated based on the results of computer climate models. In Chapter 3 we look at the implications of such policies.

Governmental action should be taken only when scientific theories are substantiated more concretely than has been done in the case of global warming. If you read this book with an open mind, you will see that the documented evidence suggests that global warming caused by man's activities is so far unsubstantiated.

Conservation of our natural resources is vital to sustaining human life and society on this planet. Prudent control of our ever-increasing population is a critical task. In the long term, overpopulation, not some imperceptible change in the earth's temperature, will cause the most damage to the human species and its environment.

The earth's temperature has fluctuated throughout geologic time. Species have come and gone. Those species that have adapted are the ones that have survived. Humans have proven to be among the most adaptable of all the species, and we will continue to adapt and survive.

REFERENCES

1. Svante Arrhenius, *Worlds in the Making*, trans. H. Borns (New York: Harper & Brothers Publishers, 1908), p. 63.
2. G. N. Plass, "The Carbon Dioxide Theory of Climatic Change," *Tellus* 8 (1956):142.
3. Ibid., 141–154.
4. Charles D. Keeling, "The Concentration and Isotopic Abundances of Atmospheric Carbon Dioxide in Rural Areas," *Geochim. et Cosmochim. Acta* 13 (1958):322–334.
5. Syukuro Manabe and Richard T. Wetherald, "Thermal Equilibrium of the Atmosphere with a Given Distribution of Relative Humidity," *J. Atmospheric Sci.* 24 (1967) 241.
6. Richard R. Kerr, "Hansen vs. the World on the Greenhouse Threat," *Science* 244 (1989):1041.
7. Ibid., 1043.
8. Ibid., 1042.
9. Jose P. Peixoto and Abraham H. Oort, *Physics of Climate* (New York: American Institute of Physics, 1992), p. 118.
10. J. T. Houghton, G. J. Jenkins, and J. J. Ephraums, eds., *Climate Change: The IPCC Scientific Assessment* (Cambridge: Cambridge University Press, 1990), p. 48.
11. J. T. Houghton, B. A. Callander, and S. K. Varney, eds., *Climate Change 1992* (Cambridge: Cambridge University Press, 1992), p. 5.
12. Ibid., 111.
13. William H. Calvin, *How the Shaman Stole the Moon* (New York: Bantam Books, 1991), p. 26.
14. Syukuro Manabe and Ronald J. Stouffer, "Century-Scale Effects of Increased Atmospheric CO_2 on the Ocean-Atmosphere System," *Nature* 364 (1993):215–217.
15. Christopher Anderson, "Tales of the Coming Mega-Greenhouse," *Science* 261 (1993):553; Andrew J. Weaver, "The Oceans and Global Warming," *Nature* 364 (1993):192–193; "Looking Far Ahead into the Greenhouse," *Science News*, 14 August 1993:111.

TWO

Environmental Hysteria

Humanity is headed for a fiery furnace of self-imposed doom. The problem: Global warming. The cause: Our emissions of carbon dioxide and certain other gases. Left unchecked, these emissions will alter the Earth's climate so rapidly and so thoroughly that the sea will rise, destroying coastlines and inundating islands. Entire species of plants and animals will be exterminated, and civilization will be irrevocably changed.

—GLOBAL WARMING MYTH

The global warming myth, as described in Chapter 1, is often so overstated that it can only be characterized as hysteria. Predictions of dire consequences resulting from the doubling of atmospheric carbon dioxide have prompted gloom-and-doom scenarios in environmental awareness books, textbooks, government reports, and newspaper and magazine articles. Most of these publications, while purporting to be objective, begin with one-sided, often apocalyptic, views of a world in imminent danger. Though they often go on to provide a somewhat objective presentation of the issue, they have nonetheless perpetuated the hysteria, because the first impression is the lasting impression.

At the other end of the spectrum are the debunkers who rail at the very idea that humans could destroy the earth. While most of the books and articles by these extremists provide some balance to the global warming myth, their titles, such as *Apocalypse Not* and *Eco-Scam*, scream "It's all nonsense!" They prey on our fears of excessive government control and of extreme environmentalism, and they provide just the ammunition needed by the superconservative army, shouting the battle cry "Conspiracy! Scam!"

Caught in the middle are the scientists, the schools, the public, and perhaps most influential, the politicians. Who's right? Where does the hysteria come from? And what harm does it do?

THE MYTH OF TODAY

Let us look at some examples of hysteria in the environmental books of the 1980s and 1990s. Many of these books were written by respected authors, including scientists with considerable credentials. The assumptions of the global warming myth, including the cause-and-effect link between carbon dioxide and atmospheric temperature, appear in all these books. They are accepted as fact, and yet, as we shall see in later chapters, this issue has not been resolved scientifically.

The prologue to *Dead Heat*, written by Environmental Defense Fund scientist Michael Oppenheimer and writer Robert H. Boyle, begins with this passage:

> Humanity is hurtling toward a precipice. Left unchecked, the emissions of various gases, particularly carbon dioxide from fossil-fuel combustion and deforestation, are likely to alter the Earth's climate so rapidly and so thoroughly as to destroy much of the natural world and turn civilization upside down.[1]

After six pages of bleak warnings, the authors admit the uncertainties involved in making such projections and spend much of the book criticizing computer climate models!

Greenhouse Earth is a book written to explain to the layperson the report of the Intergovernmental Panel on Climate Change (IPCC), the international organization established in 1988 to coordinate the work being done worldwide on climate change issues. The book begins with an imaginary letter from the future that describes life for children in two parts of the world where the effects of doubled carbon dioxide have created extreme climate shifts, causing hardships and unhappiness.[2] While the book goes on to give an insightful analysis of the panel's report, pointing out the uncertainties of computer models and global warming predictions, it has nonetheless perpetuated the myth of impending disaster.

The preface of another book, *The Challenge of Global Warming*, begins:

> Fossil fuel burning, deforestation, and the release of industrial chemicals are rapidly heating the earth to temperatures not experienced in human memory. Limiting global heating and climatic change is the central environmental challenge of our time.[3]

The theme of environmental challenge is not new, as we shall see, but it seems that with each decade there is a different environmental issue that becomes the "central challenge."

Taking the worst-case approach, popular science writer and novelist Isaac Asimov, along with novelist Frederik Pohl, writes in *Our Angry Earth*:

> Given just the right conditions—enough burning to turn all the world's fossil fuels into smoke, enough warming to release all that stored carbon in tundras and sea waters and turn it into carbon dioxide—there is a real possibility that the global warming would become a runaway, feeding on itself, and never stopping until there was no longer enough free carbon and oxygen left in the world to react.[4]

Most of the book rests on the assumption that the "right conditions" will occur and that drastic action is needed. The authors recommend such things as boycotts, letter-writing campaigns, group activism, and political involvement.

One of the best-known climatologist-writers is Stephen Schneider, of the National Center for Atmospheric Research in Boulder, Colorado. Schneider's 1989 book *Global Warming: Are We Entering the Greenhouse Century?* provides a complete description of the global warming issue and is very candid about the uncertainties. However, Schneider presents an apocalyptic scenario of the future, and despite his assurance that the description is fictional, he is very convincing with the underlying conviction that the events will occur if nothing is done.[5] Schneider consistently recommends drastic cutbacks in carbon dioxide emissions in all of his publications and speeches, always warning that it is better to be safe than sorry.

Perhaps one of the more influential books on the subject is *Global Warming: The Greenpeace Report*. Greenpeace is an environmental organization that takes a dramatic approach to environmental issues, and it considers global warming to be one of the greatest dangers facing the world. When the IPCC published its 1990 report *Climate Change,* Greenpeace criticized the report, accusing the panel of failing in its responsibility to recommend drastic cuts in carbon dioxide emissions. *The Greenpeace Report* was compiled to explain the IPCC report and reinterpret its conclusions. According to the report's editor, Jeremy Leggett, *The Greenpeace Report* says what the IPCC *should* have said.[6] Not everyone agrees that the IPCC should have been more extreme in its recommendations, but *The Greenpeace Report* has had a strong impact on writers, reporters, and policymakers.

These books and others have helped to mold the attitude of the public and have lent credence to the view that dangerous warming is an undeniable probability.

RIVALING MYTHS: REMEMBER THE COOLING?

The hysterical statements about global warming that assault us now are reminiscent of those of the 1970s, when the warnings were about global cooling. Scientists looking at temperature data from the 1930s to the early 1970s noted that the world seemed to be cooling down, and some of them announced that we were about to enter the next ice age. One of those scientists, now a leader in the global warming movement, was Stephen Schneider. In 1971 Schneider predicted in the journal *Science* that, because of air pollutants from human activity, a decrease in global temperature by as much as 3.5°C (6.3°F) could be expected in the next fifty years. According to Schneider, this decrease sustained over just a few years would be sufficient to trigger an ice age.[7] Schneider's warning was taken seriously, and many books and articles on the subject followed.

One of those books was *The Cooling*, in which the author, Lowell Ponte, a science writer, warned that the earth faced a great ice age in less than 100 years. "It is a cold fact: the global cooling presents humankind with the most important social, political, and adaptive challenge we have had to deal with for ten thousand years."[8] Sound familiar? (Compare this statement with the previously mentioned preface of *The Challenge of Global Warming*, in which global heating is called the "central environmental challenge of our time.") Ponte's position was influenced in part by a 1975 report from the National Academy of Sciences that suggested the possibility of such cooling, based on temperature statistics from 1880 to the 1970s. (By 1991, the Academy's stance had changed dramatically, and in its 1992 report, *Policy Implications of Greenhouse Warming*, the focus was on the issue of warming, its potential effects, and recommendations for action.)

In the Foreword to *The Cooling*, Reid Bryson, a climatologist at the University of Wisconsin, told us that Mr. Ponte had "put the map of climatic arguments into a reasonable perspective,"[9] clearly agreeing with Ponte's position. Dr. Bryson was one of those who believed that global cooling was a serious potential problem. He is still concerned about climate, but he has not jumped on the warming wagon like some of the others who have drastically changed their positions. In fact, in 1993 he wrote:

> Using the only model that has demonstrated ability to simulate the five-years' mean variations of climate in the past century . . . the conclusion is reached that, when carbon dioxide doubles, the Northern Hemisphere temperature will be about what it was in 1951. That hardly seems catastrophic as we got through 1951 without much difficulty.[10]

Dr. Bryson's research on atmospheric particulates from volcanoes, a climatic influence we discuss in Chapter 9, reinforces his position that global warming is not the "central issue" we should be worrying about.

In the frenzy of this fear of freezing, scientists made suggestions for schemes that would modify the climate: Cover the Arctic ice cap with dark soot that would absorb sunlight and melt the ice in only three years; dam the Bering Strait and pump water from the Arctic Ocean into the Pacific, drawing water from the Atlantic in from the other side and raising the temperature enough to melt the ice; use nuclear devices to fragment the ice and stir up the warmer water from below; divert the arctic's rivers. Stephen Schneider discussed some of these in his 1976 book, *The Genesis Strategy*, pointing out the possibility of inadvertent adverse consequences of these remedies, such as climate changes in other areas, sea level changes, and radioactive contamination.[11] (By this time Schneider was predicting warming while others were still worried about cooling.) Clearly, the complexities of the earth's systems need to be understood better before drastic action is taken to alter them.

Today's suggestions for modifying the climate have the same ring of man's arrogance. There is even a name for this field of atmospheric tampering: geoengineering (which we look at in further detail in the next chapter). Some geoengineering solutions discussed in the 1992 National Academy of Sciences report include injecting dust or soot into the atmosphere, seeding clouds with particulates, placing mirrors in space, and fertilizing the ocean's algae. As with the solutions of the seventies, some of these proposals carry with them the adverse consequences of ozone depletion, acid rain, and damage to the ocean's ecosystem.[12] It is interesting to note that Schneider mentioned some of these possibilities in *The Genesis Strategy*: "In a more fanciful vein, why not scatter even larger particles (or mirrors) in orbit around the earth . . . ?"[13] Apparently he did not take these proposals too seriously!

Amid the global warming hoopla, there are still voices crying out about an impending ice age and demanding immediate action to prevent it. John Hamaker, an engineer-turned-farmer who became concerned about soil health, developed the theory that a sustained greenhouse effect brought on by increased carbon dioxide could be the trigger for another ice age. He has asked for some rather startling actions. His plan calls for, among other things, spreading billions of tons of gravel dust on forests and jungles, plus more dust on millions of square miles of land where forest *could* grow—"rock around the clock" is how Larry Ephron, one of Hamaker's disciples, describes the project in *The End: The Imminent Ice Age & How We Can Stop It*. (All construction on the planet would need to be halted so that cement plant grinders could be used exclusively for producing finely ground gravel dust.) In addition, Hamaker suggested that sludge left from city sewage treatment could be mixed with the gravel dust, and the microorganisms from the sewage would help clean up our rivers and harbors, using no toxic chemicals. The project would only cost about a trillion dollars.[14] Imagine all the countries on earth agreeing to such a project! Fortunately, little attention has been paid to Hamaker's suggestions.

And so we see that environmental hysteria is not new. The fears of a new ice age, the warnings of global warming, the threat of an ozone hole are all examples of the environmental "crisis of the day." Remember Aesop's fable of the boy who cried wolf? The townspeople responded to his false cry several times, but by the time the wolf actually attacked, the people had become jaded and ignored the boy's shouts. Will the public be jaded to today's environmental crises and ignore the warnings if a threat proves to be real?

MYTH-INFORMATION

While the global warming myth contains some facts, it also contains exaggerations and actual misstatements by some scientists and environmental groups. These statements are reported in the media and fuel the fire of hysteria. Let us look at some common statements that appear in articles and books and compare them with information backed up by research. (Subsequent chapters will examine these facts and assumptions in greater depth.)

MYTH: The temperature of the earth has increased approximately 0.5°C (0.9°F) in the last 100 years because of human activities.

INFORMATION: There are several problems with this statement. First of all, scientists are not sure whether there has been a notable increase. According to the 1990 report of the IPCC, the mean global temperature has increased by 0.3–0.6°C (0.5–1.1°F) over the last 100 years, but the report states that this increase is no greater than that caused by natural climate variability.[15] This report is a major assessment of current research and is used as an authoritative source of information by scientists, writers, environmental groups, and policymakers.

In fact, there may not have been an increase at all. According to Susan C. Whitmore, of the United States Department of Agriculture, a group of MIT climatologists studied worldwide ocean temperatures since 1850 and found little or no indication of global warming over the past century.[16] Furthermore, there is considerable inconsistency in the temperature data gathered during the last hundred years, as we shall see in Chapter 5. So the reported 0.5°C temperature increase is not even verified!

MYTH: Of the greenhouse gases, carbon dioxide is the gas doing the most damage. If it continues being poured into the air at its present rate, CO_2 will be responsible for about half of the greenhouse-effect warming by the year 2020.

INFORMATION: The relationship between carbon dioxide and atmospheric temperature is not yet fully understood. Some ice core and tree ring studies of temperature trends and carbon dioxide indicate that our climate has undergone substantial warming periods in the past without increased levels of carbon dioxide, and they show that increased carbon dioxide is not the only factor that might cause warming. The major greenhouse gas culprit is probably water vapor (Chapter 8). Furthermore, there is still a question of which came first, the increase in carbon dioxide or the rise in temperature. Chapter 6 provides a discussion of this puzzle.

Another factor that enters the picture is the Mystery of the Missing Carbon. There is a large discrepancy between the amount of carbon dioxide being emitted and the amount measured in the atmosphere. What is happening to it? (Read Chapter 6 for some clues.)

MYTH: Global temperature will increase another 1.5–4.5°C (2.7–8.1°F) by some time in the next century if we do not take drastic measures now. (In the more apocalyptic scenarios, only the extreme prediction of 8°F is used.)

INFORMATION: Recent data from satellites showing little or no recent change in global temperatures suggest that there is time to learn more and to benefit from advancing technology. The lack of confidence in current computer predictions should dictate that drastic action be postponed until further knowledge is gained.

MYTH: Global warming will cause the polar ice to melt, raising sea levels all over the globe, flooding entire regions, destroying crops, and displacing millions of people.

INFORMATION: There is still a great deal of uncertainty regarding a potential rise in sea levels. If the earth warms, sea levels will rise as the water heats up and expands. If the polar ice caps melt, more water will be added to the oceans, raising sea levels. There is some evidence that melting has occurred, but there is also evidence that the Antarctic ice sheets are growing. In fact, it is possible that a warmer sea-surface temperature will cause more water to evaporate, and when wind carries the moisture-laden air over the land, it will precipitate out as snow, causing the ice sheets to grow. Certainly, we need to have better knowledge about the hydrological cycle (Chapter 8) before predicting such dire consequences of higher temperatures.

The estimate of the expected rise in sea level has lowered continually since global warming was first predicted. William Nierenberg, of the Scripps Institution of Oceanography, expressed his concerns about such predictions in an article in the journal *Environmental Conservation*:

> It was only some 17 years ago that serious predictions of a 25 ft. rise in a matter of "decades" was made by an expert. This estimate has been steadily reduced. . . . In fact, there are some who believe, on the basis of actual measurements of the change in Greenland ice-cap thickness and model predictions, that the average sea-level may even decrease! . . . The current version of the IPCC barely mentions sea-level rise.[17]

Even so, sea level rise is one of the consequences of global warming most commonly mentioned in textbooks and news articles.

MYTH: Human society's effect on the planet matches the size and impact of natural processes.

INFORMATION: While there might be some climate effects from human activity, calling human activity the "major cause" is an overstatement. Natural events appear to have the greatest effect.

For instance, as we discuss in Chapter 9, the 1991 eruption of Mt. Pinatubo in the Philippines had a definite effect on temperatures, causing a decrease in the average global temperature. And while the cooling effect may be nearing its end, it represents the dramatic, yet unpredictable, effects of nonhuman influences on global climate change.

Another natural influence is El Niño, a periodic perturbation in the ocean's temperature and circulation that causes extreme global climatic events, including droughts and major flooding.

Of even greater importance to the earth's climate are variations in the sun's radiation, covered in Chapter 7, and variations in the earth's orbit, a phenomenon that is explained in Chapter 5.

The truth is, climate variability has always existed and will continue to do so, regardless of human intervention.

WHENCE COMETH THE PROPHETS OF DOOM?

Who is responsible for the hysteria, and why would anyone want to cause it? Let us look at the sources and possible motivations.

Media Watch

The news media, in one form or another, reaches most of the population. Because of this large audience, reporters and commentators have a great influence on public opinion and politics. Dixy Lee Ray, an outspoken critic who blamed most of the environmental hysteria on the news media, was a scientist who spent her later years chastising the environmental movement. She had a lot of experience with environmentalists, having been chair of the United States Atomic Energy Commission (now the Department of Energy, or DOE). She was also a professor of zoology at the University of Washington, as well as a former governor of the state of Washington. In her 1993 book *Environmental Overkill*, Dr. Ray used a discussion of the 1992 Earth Summit to show how the media shaped the public's impressions of the proceedings, particularly of the treaties. The controversial biodiversity treaty was described as "designed to protect plants and animals" when actually the text, which was seldom mentioned, demanded that the United States provide foreign aid to Third World countries with no conditions![18] Most people consider plant and animal protection to be a noble goal, but how many would favor unrestricted financial aid to developing countries? And yet, this detail was carefully avoided in news reports, creating a negative response to President Bush by environmental groups and citizens of the United States and other countries.

Who can blame the media? After all, disaster is what sells. Screaming headlines get attention, but does everyone read the entire article or story? The tabloids at the grocery checkout counter are expected to exaggerate, but consider these headlines from mainstream publications:

- "Preparing for the Worst: If the Sun Turns Killer and the Well Runs Dry, How Will Humanity Cope?" (*Time*, 2 January 1989)
- "Our Fragile Earth: Only Now Are We Beginning to Fathom the Gravity of Our Environmental Dilemma" (*Discover*, October 1989)
- "The Heat Is On: Chemical Wastes Spewed into the Air Threaten the Earth's Climate" (*Time*, 19 October 1987)
- "Icy Indicators of Global Warming" (*World Watch*, January–February 1993)
- "Global Warming May Spur Climatic Chaos, Scientists Say" (*Los Angeles Times*, 15 July 1993)
- "No Way to Cool the Ultimate Greenhouse" (*Science*, 29 October 1993)

- "Gore Labels Global Warming Top Peril" (*Los Angeles Times*, 22 April 1994)

News stories usually include a summary of the global warming issue, using the popular view. Occasionally they mention—usually toward the end—that not all scientists agree. But often, the attention-getting title misrepresents the text of the article. Consider this title that appeared in *The New York Times* on January 17, 1995: "Most Precise Gauge Yet Points to Global Warming." The article, written by Malcolm W. Browne, describes two years of satellite sea level data presented at a meeting of the American Geophysical Union, data that indicate a 3-millimeter rise in global sea level (3 millimeters equals about one-tenth of an inch). According to the article, participants in the project "acknowledge that two years of observations cannot prove the existence of long-term climate trends." The period reported on does not include a year with the El Niño effect. According to Dr. R. Steven Nerem, of NASA's Goddard Space Flight Center in Greenbelt, Maryland, including El Niño episodes will probably cause the rise to be only 1 to 2 millimeters. And yet Browne states near the beginning of the article that if the seas continue to rise, "a time will come when entire countries, Bangladesh and the Netherlands among them, are inundated."[19] There are no figures given for the amount of sea level rise needed for such dire consequences. This perceived rise is in fact even less than the 1992 IPCC predictions.

Unfortunately, many people read only the titles and beginnings of news articles and this type of reporting leaves them with the wrong impression. Where is the line drawn between reporting the news and making the news?

A Textbook Case

What role does the educational system play in the current hysteria? A look at textbooks and curriculum enrichment materials shows that the system is often firmly in the hands of the doomsayers.

When we conducted a survey of textbooks and supplemental materials that include discussion of the greenhouse effect, we found that over half of them presented the extreme view, with little or no disclaimer about inconsistencies. The others took a somewhat balanced approach.

Here is an example: A college textbook (used in a graduate-level environmental studies course) entitled *Environmental Science* includes these passages:

> Although we have been on the earth for only an eye blink of its existence, we are now altering the chemical content of the earth's atmosphere 10 to 100 times faster than its natural rate of change. Projected global warming caused by our binge of fossil fuel burning and deforestation, ozone depletion caused by our extensive use of chlorofluorocarbons and other chemicals we could learn to do without, and the buildup of nuclear weapons are now major global environmental threats.[20]

No citations. Sweeping statements. Judgmental declarations. To continue this poetic but unscholarly approach, the book goes on to say that "if we keep pumping greenhouse gases into the atmosphere and continue cutting down much of the world's forests, we are flipping a coin and gambling with life as we know it on this planet."[21] This seems extreme and subjective for a scientific textbook.

"Race to Save the Planet," a popular video series produced by WGBH for public television, features actress Meryl Streep as the hostess-narrator. In it, the standard litany of predictions is recited: 8°F temperature increase, sea level rise, droughts, deforestation. There is no disclaimer about the uncertainties in these predictions, no mention that the global warming theory is just theory, based on computer models, and that not all scientists agree. The series is used in many secondary school and university science and environmental classes.

Some of the supplemental material used at the middle school level uses extreme language and dire predictions, and it frightens and depresses many children. What purpose does this serve? Without the benefit of a balanced presentation of the issue, these youngsters are left with a sense of futility about their future.

Deep Ecology

Environmental organizations are at the frontlines of the wave of hysteria. Eco-warriors, the radical fringe of the environmental movement, sometimes called eco-freaks, have a serious interest in perpetuating fear. Fear is a powerful aid in mobilizing forces to back one's cause.

The environmental movement of this century caught the public's interest in the seventies, a period of consciousness-raising among the under-thirty crowd. Interest in conservation and ecology grew at a remarkable rate and led to the formation of many organizations dedicated to solving environmental problems. According to Donald Snow, of the Conservation Fund, there are now some 10,000 conservation and environmental organizations, and they range in size from small community clubs with voluntary staffs to national institutions with multimillion-dollar budgets, paid staffs, and hundreds of thousands of members.[22] Many of these groups do a remarkable job of protecting and preserving our natural resources. Some, however, such as the extreme EarthFirst! movement, not only lose sight of common sense, they undermine rational efforts to solve problems.

The early environmentalists, such as John Muir and Gilbert Pinchot, sought to protect the environment for man's enjoyment. The movement has changed, and even the less radical groups like the Sierra Club lean toward protecting the environment *from* man rather than *for* man.

Radical environmentalism, or deep ecology, can be credited with much of the momentum of environmental hysteria. Edward Abbey, one of the forefathers of deep ecology, clearly believed that society was on the wrong track. In *Green Rage: Radical Environmentalism and the Unmaking of Society*, Christopher Manes, himself a radical environmentalist, presents Abbey's view that if "we can draw the line against the industrial machine in America, and make it hold, then perhaps in the decades to come we can gradually force industrialism underground, where it belongs."[23] Abbey inspired many of today's radicals with his 1975 novel *The Monkeywrench*

Gang, which tells of the adventures of a group of societal dropouts who set out to sabotage the development and industrialization of the American West. In a short time, the Monkeywrench Gang became the very real EarthFirst! gang, a group that has become known for spiking trees, burning bulldozers, and generally attempting to bring attention to what they consider to be a defective American dream. In fact, in Abbey's last novel, *Hayduke Lives*, he described EarthFirst!, even using the organization's name. Let's listen in on a conversation:

> ". . . not only does the eco-warrior work without hope of fame and praise, not only does he work in the dark of night amidst a storm of official public calumny, but he works without hope of pecuniary recompense."
>
> "What do you mean no pecuniary recompense?" said George. "Hell's f___, Doc, us terrorists got to live too."
>
> "True, but only on a subsistence level. We want no mercenaries in the ranks of our eco-warriors. As I said, you do your needed work out of love, the love that dare not speak its name . . ."[24]

Later:

> "Don't forget Rule Number Two," Doc said, opening his beer.
>
> "Don't hurt anybody." Hayduke had already opened another himself. "Murder only in self-defense."
>
> "No no no, that's Rule Number One. Rule Number Two is, Don't get caught, remember?"
>
> "I know, I know, got to clean out that trunk one of these days." He guzzled his beer.[25]

If you have any doubts about the zeal of eco-warriors, read these two novels!

Today's agenda is clear. In *Eco-Warriors*, another look at radical environmentalism, Rik Scarce describes the views of today's activists:

> "The time for compromise by the environmental movement has long since passed," said David Brower, former executive director of the Sierra Club and one of those most responsible for today's environmental movement. "A pluralistic society must compromise, but the compromise must be between advocates, not compromisers. The public has grown tired of the lethargic responses by government and business to the eco-catastrophes now here."[26]

Well, perhaps. But it is just as likely that the public will grow tired of worrying about overblown catastrophes and crises du jour. When and if the wolf comes, the villagers will have long since stopped listening to the cry of the boy.

While the Edward Abbeys and the David Browers of the world are idealists, however misled, what about the environmental movement as a whole? How much of the hype stems from pure concern for the world? Could it be that, as Dixy Lee Ray has said, "professional environmentalists and others whose jobs depend on the continuing environmental crises want us to think that all's wrong with the world"?[27] A revealing headline appeared in *The Washington Post* in June 1989: "Environmentalists Hope for Scorcher; Aim Is to Avert Governmental Complacency on 'Greenhouse Effect.'" The article quotes Dan Becker, director of the Sierra Club's global warming campaign: "If this summer is especially bad, a crisis mentality will take over and Congress will want to pass legislation to show that they're on top of the situation." Irving Mintzer of the World Resources Institute agreed: "If we have another hot summer, it might not mean that the greenhouse effect is here, but it will galvanize political opinion around the issue."[28] (It was the previous summer of 1988, that infamous summer of heat and fires and hurricanes, that had set the stage for Jim Hansen's dramatic proclamation that global warming had arrived.) While the Sierra Club and the World Resources Institute are more or less moderate in their approach, unlike the radical EarthFirst!, let us not forget that they are huge organizations with hungry budgets.

Science: Funding the Research

Many scientists, for completely different reasons, fuel the fire of hysteria with "data-free speculation," and they have a trusting audience. In the past, it was the scientists who called for calm in the midst of chaos; today many of the scientists themselves are sounding the alarm. Why the shift?

One reason is that career pressures have become very strong in contemporary science. Investigating fraud in science led science writers William Broad and Nicholas Wade to publish *Betrayers of the Truth* in 1982. What they discovered tells us a lot about these pressures. They point out that "the system rewards the appearance of success as well as genuine achievement. Universities may award tenure simply on the quantity of a researcher's publications, without considering their quality." They go on to say that those "who falsify scientific data probably start and succeed with the much lesser crime of improving upon existing results."[29] There is a fine line between adjusting data and falsifying it, and the line is easily crossed.

This issue of "fitting the data" is problematic in climate model science. Michael Oppenheimer describes this aspect in *Dead Heat*, admitting that scientists occasionally adjust the guesses in the models, which are very crude, so that the predictions of the model agree with the measurements—in other words, so that the model gives the "right" answer. He borrows an analogy from economists, whose methods are similar: "Models are like sausages: you don't want to know what goes into them."[30] The issue of adjusting data in computer models is examined in depth in Chapter 4.

Most scientific research is funded by federal grants from agencies such as the Environmental Protection Agency, National Science Foundation, Department of Energy, and others, and competition for available funds can be intense.

Reid Bryson, one of the climatologists who warned of cooling in the 1970s, referred to the need that scientists have for research money in a sarcastic comment; "Then an important discovery was made. It was found that large amounts of research funding could be generated by worrying about the dangers of a changing climate."[31] An event in the summer of 1994 illustrates this point. The world watched in awe as Comet Shoemaker–Levy 9 struck Jupiter, and fear of a similar occurrence on our own planet offered a splendid opportunity for NASA funding. A $50-million proposal was presented to Congress to create a comet-watch program. Several questions come to mind: If scientists were able to predict, observe, and photograph the Jupiter collision, are not the equipment and technology already in place for comet watching? And what are we going to do in the remote chance that such an event becomes imminent? Evacuate the earth?

A major discovery or publication may put undue pressure on a scientist to follow up with more major work. A physics professor at the University of Houston learned this the hard way. In "Trapped in His Own Shadow," *Los Angeles Times* science writer Mark Stein tells how Paul Chu "dazzled the science world" with his discovery of a superconducting material. Fame brought enticing offers from prestigious universities, but Chu, loyal to Houston because of their support, stayed to direct the $22.3-million research complex built to keep him there. So far he has made no further spectacular discoveries, and, with speeches to make and funding to acquire, Chu lives under tremendous pressure. Both his research and his personal life have suffered.[32] It is easy to see how the pressure to justify and finance one's research could cause one to overemphasize the dangers of a possible catastrophe.

Science and Politics—Strange Bedfellows

The marriage of science and politics is a phenomenon of the twentieth century. In earlier centuries, scientists were often wealthy people who could afford to pursue their love of learning. They did not rely on financing from anyone else. Because most of today's scientific funding comes from government agencies, science has become politicized. As reporter Ronald Bailey says in his book *Eco-Scam*:

> Lab directors are not only scientists; they are also public relations officers and politicians who must navigate the dark byways of Congress and government agencies in search of the wherewithal to keep their organizations going. Consequently, they feel enormous institutional pressure to hype the work of their laboratories and to tie it to the solution of some mediagenic crisis.[33]

In other words, scientists must do more than legitimate scientific research. They have to know how politicians think, and they have to create a need for the research on which their livelihood depends.

How does a scientist create this need? In an interview published in *Discover*, Stephen Schneider said, "we have to offer up scary scenarios, make simplified, dramatic statements, and make little mention of any doubts we might have."[34] Schneider repeatedly points out in his writing that there is great uncertainty in computer climate models. But he is a pragmatist; uncertainties do not inspire research grants.

One of the most obvious examples of this marriage of science and politics is the global warming issue. As Sonja A. Boehmer-Christiansen points out in a 1994 *Nature* commentary, "global warming could not have entered international politics without the support of influential voices from the scientific community." She questions why and how scientists publicized a concern that was strictly in the research stages. This unsubstantiated theory quickly led to an international treaty with strong economic and political ramifications. Boehmer-Christiansen is straightforward in her opinion of why scientists jumped on the global warming bandwagon: "Under pressure, even scientists will deliver what their paymasters prefer to hear."[35] It is unfortunate that science has come to this, that the pressures of research funding can have such a profound political effect.

Another way in which science and politics become closely intertwined is exemplified in the strong environmental activism of individual politicians. Consider Vice President Al Gore's book *Earth in the Balance*.[36] The book takes a scholarly approach to environmental issues, but it comes down heavily on any detractors and calls for Draconian actions. Mr. Gore was heavily influenced in his environmental leanings by one of his college professors, Roger Revelle. Revelle was involved in early carbon dioxide monitoring and was part of the original group of climatologists warning of global warming. Interestingly enough, Revelle backed off from the extremists, coauthoring a paper with Chauncey Starr and S. Fred Singer (we will talk about him in the next section) that suggested that the uncertainties in the global warming theory did not justify drastic reductions in greenhouse gas emissions. There has been quite a controversy about the article, which was published just before Revelle died, and accusations were made by one of Revelle's former students that Singer took advantage of Revelle's advanced age and mental state. Singer sued for libel, and eventually the student retracted his statements.

As second-in-command of one of the most influential countries in the world, Gore has a great deal of potential power. Later we'll talk about his diatribes against those who question the myth.

IS THE WOLF AT THE DOOR?

Government reports, textbooks, and newspaper articles frequently announce that "most scientists" agree that global warming is a problem that is exacerbated by humans and can only be solved by radical means. Is there really a consensus among scientists?

No, there is not. While many famous scientists have signed petitions and given their nod of approval to the calls for action, not all scientists agree with the theories, and many of them are concerned about the hysteria that is being generated. We have already mentioned Reid Bryson, who has considerable evidence that volcanic eruptions greatly influence global climate. Another of the longtime detractors is Sherwood Idso, of the United States Water Conservation Laboratory in Phoenix, Arizona. In 1980 Idso upset the climate modeling community when he published an article in *Science* criticizing the approach of the climate modelers, and he followed up the next year with an article in *New Scientist*. In these articles he described how his research, using the empirical approach, differed from that of the modelers, who work with numerical theory. The empirical approach measures the response of the real world to a natural event, while computer models calculate responses by means of equations. Idso's results from twelve years of field experiments, which we discuss in Chapter 6, showed that an increase in carbon dioxide would actually be desirable because of its helpful effects on plant growth. With such a major discrepancy between computer projections of the detrimental effects of carbon dioxide on climate and his real-life experiments showing positive benefits, he warned that action should be taken cautiously.[37] In other words, we should not put our efforts into eliminating potentially beneficial greenhouse gases until we know for sure whether they are having a harmful effect.

Dr. Idso continues his research, and in a telephone conversation in April 1994, he enthusiastically discussed his ongoing CO_2 experiments, which hold the promise of a greener, healthier climate. Perhaps man's accidental interference has in reality provided, as he says, a "breath of fresh air for the planet's close-to-suffocating vegetation,"[38] a benefit that would be lost if drastic reductions in carbon dioxide emissions were made.

Robert C. Balling, Jr., a geography professor at Arizona State University and director of the Office of Climatology, agrees that an increase in atmospheric carbon dioxide might be beneficial. Balling has done considerable climate research, and in his 1992 book, *The Heated Debate: Greenhouse Predictions Versus Climate Reality*, he questions the global warming myth as it is currently understood, providing evidence from his research that contradicts the predictions of the global warming enthusiasts. Balling uses the term "greenhouse gospel" as he points out the obvious exaggeration and seriously flawed predictions, and he believes that any warming that might occur will be moderate and possibly beneficial.[39] Balling does not advocate ignoring the issue because it is unproven, just as we do not, but he shares our view that the public should be informed.

A prominent climatologist who does not share in the "consensus" is S. Fred Singer, director of the Science and Environmental Policy Project, an independent research and education institute. Singer was at one time the director of the National Weather Satellite Center and was also a professor of environmental sciences at the University of Virginia. He says: "There is . . . disagreement in the scientific community about predicted changes as a result of further increases in greenhouse gases; the models used to calculate future climate are not yet good enough." He goes on to point out that "panicky, premature" steps (cutting carbon dioxide emissions) to delay an as-yet uncertain warming would be economically devastating (to the tune of a few trillion dollars) without being effective.[40] While many climate modelers express similar doubts (see Chapter

4), the government plunges ahead making policy, reacting to the predictions of doomsayers, and shunning the cautionary pleas for greater scientific certainty.

THE RISKS OF OVERREACTING

What are the dangers of overreacting, besides disrupting the economy? Think again of the story of the boy who cried wolf. For every cry of wolf there is a response. The immediate rush to arms, laying down the work of the day, undermines the longer-term ability to respond. Fear of the unseen wolf overcomes objectivity. And the strident repeated cries deaden sensitivity to real issues and create a backlash.

Backlash

Any radical movement creates a backlash. The extent to which the backlash affects public opinion and governmental policy depends in part on the direction of the backlash. Because the global warming scenario calls for controlling industrial gas emissions, both industry and extreme conservatives have launched the backlash. The industrial economy will be substantially affected by emission controls, while the conservative movement sees government restrictions as a threat to personal liberty.

Much of the backlash comes from conservatives such as Rush Limbaugh, a popular talk-show host. When Limbaugh, with his wide following and substantial public influence, declares that there is no global warming and assures us that we have a right to use the earth's resources indiscriminately to make our lives better, people listen. He convinces many of his followers that the "environmental wackos" are out to destroy our way of life.[41] Despite his flawed science, his confrontational manner, and his stereotyping, he makes sense to many people. In fact, we sometimes find ourselves agreeing with him; he makes many of the same points we make. But Limbaugh carries his criticism of environmentalists to such an extreme that he convinces his listeners that all environmentalism is wrong.

Do the debunkers of environmentalism have the answers? Of course not. Absolving ourselves of environmental guilt is no more responsible or effective than creating environmental panic. The extreme backlash encourages the activities that lead to depletion of limited resources, pollution of the air, water, and landfills, and general apathy toward the future of the earth.

Loss of Objectivity

Another casualty of overreacting is objectivity. The pressure to conform to the "correct" view creates a stifling atmosphere. Those who question the conventional wisdom run the risk of favoring industry, of endangering the environment, of naïveté, or worse, of ignorance.

Whether for political reasons or not, those who question the commonly accepted theories are in danger, and this is another frightening dimension of the marriage of science and politics.

When scientific inquiry becomes a threat to a political viewpoint, what happens to objectivity? Consider the experience of William Happer, a Bush appointee to the Department of Energy who was retained by the Clinton administration. Happer was not happy with what he considered an overstatement of the health threat of ozone depletion, nor was he discreet about his opinion, and he was fired from the position.[42] It appears obvious that his questioning of the current ozone beliefs lost him his job. This story has dangerous overtones of censorship.

Objectivity is also at risk when challenging the commonly accepted theories because of the danger of being associated with the opposition. In the case of global warming, the primary enemy of environmentalists is industry, and global warming skeptics are therefore accused of favoring industry. The following incident is a perfect illustration of this problem.

In April 1994 Vice President Al Gore celebrated the twenty-fourth anniversary of the first Earth Day by declaring war on skeptics. In an article in the *Los Angeles Times,* he was reported as comparing the scientists who question the global warming theory to tobacco industry executives.[43] (This came at a time when the tobacco industry was involved in congressional hearings over whether they had misused data to downplay the negative health effects of nicotine and cigarette smoke.) The vice president is an ardent environmentalist, and he is an intelligent, well-educated man who seems credible. But has his fervor closed his mind to the importance of scientific questioning?

Fortunately, there is some rationality in the White House. While Gore was castigating the global warming skeptics, President Clinton was reminding environmentalists that "government should encourage people to work together, not pit business and workers and environmentalists against each other."[44] Clinton's attitude of cooperation with industry is not at all popular with environmentalists, who accuse the administration of "greenwashing" the Climate Change Action Plan.[45] The extreme dislike that environmentalists have for industry is evident in their closed-minded rejection of Clinton's approach. Let us hope that the atmosphere of open inquiry will not become polluted by enviro-political zealotry such as that displayed by Al Gore and other politicians.

Enervation

A more insidious danger of overreacting is that we will be so overwhelmed by problems over which we have no control that we may not be able to summon the resolve required to tackle those things we can do something about. There is also a limit to our economic resources. Throwing vast sums of money and energy at unproven enemies could leave no resources to fight other, more important battles.

What kind of resources are we talking about? According to the Department of Energy, the cost to the United States of reducing carbon dioxide emissions 20% in the next ten years could be as much as $95 billion per year.[46] Another estimate, based on information from the Office of Technology Assessment, puts the cost at between $350 billion and $520 billion per year. Taking into account fuel savings, the program could save $20 billion or cost up to $150 billion per year.[47] These and other costs are discussed in the next chapter.

Since anthropogenic carbon dioxide emissions represent only a small portion of potential global warming causes, where will we find the resources to deal with real crises that arise?

WHAT CAN YOU DO?

We have seen that environmental hysteria comes from several directions. The myth of global warming derives from predictions of computer climate models, but it is spread by scientists, environmental activists, educational systems, politicians, and by the ubiquitous news media. The backlash is influential, leaving few people with a rational, objective viewpoint.

Considering the uncertainties that remain in the global warming theory, we think it is important that the public be aware of the hysteria and its possible consequences. But what can one person do?

You need to ask questions, to keep your mind open. When you read an article or a book or see a television show, ask yourself, "What is the purpose of this report? What is the source? Does this person or organization have a vested interest in the subject?" Watch for scare tactics, exaggerations or distortions, personal attacks, and inflammatory language. Be wary of anything that "everyone" does or believes.

Before you support a political candidate or sign a petition, whether pro- or anti-environment, learn what the candidate really stands for, what the petition really demands. Doing the wrong thing for the right cause is still wrong and can have long-lasting repercussions. Do not fall into the trap of supporting legislation just because it wears a "good" label. Find out whether or not there are hidden agendas.

Be aware of the source of the myth as well as the perpetuation of it. The myth is not necessarily all false, but we as a world civilization must be sure of the facts before taking drastic action.

REFERENCES

1. Michael Oppenheimer and Robert H. Boyle, *Dead Heat—The Race against the Greenhouse Effect* (New York: Basic Books, 1990), p. 1.
2. Annika Nilsson, *Greenhouse Earth* (Chichester: John Wiley & Sons, 1992).
3. Dean Edwin Abrahamson, *The Challenge of Global Warming* (Washington, DC: Island Press, 1989), p. xi.
4. Isaac Asimov and Frederik Pohl, *Our Angry Earth* (New York: Tom Doherty Associates, 1991), p. 61.
5. Stephen Schneider, *Global Warming: Are We Entering the Greenhouse Century?* (San Francisco: Sierra Club Books, 1989).
6. Jeremy Leggett, ed., *Global Warming: The Greenpeace Report* (Oxford: Oxford University Press, 1990).
7. S. I. Rasool and S. H. Schneider, "Atmospheric Carbon Dioxide and Aerosols: Effects of Large Increases on Global Climate," *Science* 175 (1971):138–141.
8. Lowell Ponte, *The Cooling* (Englewood Cliffs, NJ: Prentice-Hall, 1976), p. xvi.
9. Ibid, xii.
10. Reid Bryson, "Simulating Past and Forecasting Future Climates," *Environmental Conservation* 20 (1993):339.
11. Stephen Schneider, *The Genesis Strategy: Climate and Global Survival* (New York: Plenum Press, 1976), pp. 205–211.
12. Committee on Science, Engineering, and Public Policy (COSEPUP), *Policy Implications of Greenhouse Warming* (Washington, DC: National Academy Press, 1992), p. 82.
13. Schneider, *Genesis*, 210.
14. Larry Ephron, *The End: The Imminent Ice Age & How We Can Stop It* (Berkeley: Celestial Arts, 1988), pp. 131–133.
15. J. T. Houghton, T. J. Jenkins, and J. J. Ephraums, eds., *Climate Change: The IPCC Scientific Assessment* (Cambridge: Cambridge University Press, 1990), p. xii.
16. Susan C. Whitmore, *Global Climate Change and Agriculture: A Summary* (Beltsville, MD: U.S. Department of Agriculture, 1992).
17. William A. Nierenberg, "Science, Policy, and International Affairs: How Wrong the Great Can Be," *Environmental Conservation* 20 (1993):195–197.
18. Dixy Lee Ray, *Environmental Overkill* (New York: Harper Perennial, 1993), p. 174.
19. Malcolm W. Browne, "Most Precise Gauge Yet Points to Global Warming," *New York Times* 20 December 1994:4.

20. G. Tyler Miller, Jr., *Environmental Science* (Belmont, CA: Wadsworth, 1992), p. 209.
21. Ibid., 215.
22. Donald Snow, *Inside the Environmental Movement* (Washington, DC: Island Press, 1992), p. xvii.
23. Christopher Manes, *Green Rage: Radical Environmentalism and the Unmaking of Civilization* (Boston: Little, Brown, 1990), p. 3.
24. Edward Abbey, *Hayduke Lives!* (Boston: Little, Brown, 1990), pp. 112–113.
25. Ibid., 113.
26. Rik Scarce, *Eco-Warriors* (Chicago: The Noble Press, 1990), p. xii.
27. Dixy Lee Ray, with Lou Guzzo, *Trashing the Planet* (Washington, DC: Regnery Gateway, 1990), p. 3.
28. William Booth, "Environmentalists Hope for Scorcher," *Washington Post* 21 June 1989:A01.
29. William Broad and Nicholas Wade, *Betrayers of the Truth* (New York: Simon and Schuster, 1982), p. 19.
30. Oppenheimer and Boyle, 54.
31. Reid Bryson, "Civilization and Rapid Climatic Change," *Environmental Conservation* 15 (1988):7.
32. Mark A. Stein, "Trapped in His Own Shadow," *Los Angeles Times* 10 July 1993:A1.
33. Ronald Bailey, *Eco-Scam: The False Prophets of Ecological Apocalypse* (New York: St. Martin's Press, 1993), p. 175.
34. Jonathan Schell, "Our Fragile Earth," *Discover* October 1989:47.
35. Sonja A. Boehmer-Christiansen, "A Scientific Agenda for Climate Policy?" *Nature* 372 (1994):400–402.
36. Al Gore, *Earth in the Balance* (Boston: Houghton Mifflin, 1992).
37. Sherwood B. Idso, "The Climatological Significance of a Doubling of Earth's Atmospheric Carbon Dioxide Concentration," *Science* 207 (1980):1462–1463; Sherwood B. Idso, "Carbon Dioxide—An Alternative View," *New Scientist* 92 (1981):444–446.
38. Sherwood Idso, *Carbon Dioxide and Global Change: Earth in Transition* (Tempe, AZ: IBR Press, 1989), p. 9.
39. Robert C. Balling, Jr., *The Heated Debate: Greenhouse Predictions versus Climate Reality* (San Francisco: Pacific Research Institute for Public Policy, 1992), p. 2.
40. S. Fred Singer, "Global Climate Change: Facts and Fiction," in *Rational Readings on Environmental Concerns*, ed. Jay H. Lehr (New York: Van Nostrand Reinhold, 1992), p. 394.
41. Rush Limbaugh, *The Way Things Ought to Be* (New York: Pocket Books, 1992), p. 168.
42. Richard Stone, "Did DOE's Happer Fall into Ozone Hole?" *Science* 260 (1993):743.

43. Melissa Healy, "Gore Labels Global Warming as Top Peril," *Los Angeles Times* 25 April 1994:A13.
44. Ibid.
45. Pamela Zurer, "Environmental Groups Fault Climate-Change Plan," *Chemical & Engineering News* 9 May 1994:28–29.
46. David E. Newton, *Global Warming* (Santa Barbara, CA: ABC-CLIO, 1993), p. 89.
47. Rosina Bierbaum and Robert M. Friedman, "The Road to Reduced Carbon Emissions," *Issues in Science and Technology* Winter 1991–1992:58.

THREE

Policy Implications and Responses

> *"Follow the middle course," warned Daedalus, as he fastened the magnificent wings of feathers and wax to the shoulders of his son, Icarus. "If you fly too close to the sun, the heat will melt the wax and you will go crashing down. If you fly too low, the sea will soak the feathers and pull you down." As father and son flew above the sea, Icarus was overcome with excitement! As he soared higher and higher, the burning heat of the sun melted the wax. Icarus flapped his arms in vain, and the sea claimed him. Daedalus, who had steadfastly followed the middle course, successfully reached land.*
>
> —MYTH OF ANCIENT CRETE

The hysteria generated over the global warming myth has led to calls for drastic governmental action. Many of the demands are extreme, with the potential for adverse economic and environmental consequences. The issue has inspired debate among scientists, economists, politicians, and environmental activists: What should the actions be, considering the many uncertainties that exist in the scientific understanding of the causes and effects, as well as the extent and direction, of climate change? Should any action be taken at all? Who should decide?

IDENTIFYING THE PROBLEM

Before considering the policy responses, let us first define the problem that government is being asked to solve. The generally accepted view of global warming holds that increasing greenhouse gases in the atmosphere, especially carbon dioxide, will lead to higher global temperatures, adversely affecting the earth's climate and thus its biodiversity, societal patterns, and land use. According to the theory, human activities contribute to most of this greenhouse gas increase, and therefore major changes must be made in these activities to reverse the trend of increasing carbon dioxide. What these changes should be and whether or not they should be made are at the root of governmental policy decisions.

Anthropogenic, or human-caused, carbon dioxide emissions come from two principal sources: burning of fossil fuels (coal, oil, gas) and changes in land-use patterns (deforestation, loss of farmland). Later we consider these sources in detail and discuss their relative effects on the atmosphere, as well as the costs of reducing emissions or continuing on the path of "business as usual."

As we have seen, the global warming theory has been developed by climate scientists, based on years of observation and research and on the recently developed computer climate models. From scientific research to government policy, especially international government policy, there is a large gap. That gap has been bridged by the environmental movement, with its activists, by the scientists who support the global warming theory, by scientific institutions that benefit from research funding, and by politicians.

The environmental movement had its first real influence on government policy when *Silent Spring*, Rachel Carson's 1962 book about pesticides, raised fears and prompted action by the government to ban DDT and other pesticides. Other books, such as Paul Ehrlich's *The Population Bomb*, warned of the dangers of overpopulation and its effect on diminishing food supplies and the fragile environment. These books raised the public's consciousness about population growth and environmental degradation, spawning environmental organizations and stimulating efforts by many people to reduce waste and conserve energy. However, the books also contributed to the fears that had been raised and fed the hysteria of overreaction. Later publications from Ehrlich and others built on these fears and escalated calls for quick, often Draconian, solutions. As we saw in Chapter 2, the issue of global warming has followed this pattern of overreaction.

In the nineties the public and political response to environmental issues is more sharply pronounced. The threat of global warming caused by anthropogenic greenhouse gases has put carbon dioxide emissions in the spotlight. There are essentially three stances taken with regard to these emissions. One is that no action should be taken to cut emissions because there is no problem. Another is that, before taking drastic action, research should continue on the possibilities of climate change, the consequences of such change, and the appropriate steps to take to affect the change or adapt to it. The last position is that, regardless of the uncertainties in our present knowledge, we should begin to take drastic action "just in case," as a sort of insurance policy.

While scientific studies, government reports, and policy statements continue to emphasize the uncertainties that exist in the computer climate models and thus in the global warming issue, the most popular approach is to proceed as if the threat is real. For example, in its 1992 report, *Policy Implications of Global Warming*, the National Academy of Sciences, a scientific society that serves as advisor to the federal government on scientific and technical matters, concluded that even with the "considerable uncertainties," global warming "poses a potential threat sufficient to merit prompt response."[1] This position reinforces the demands of environmental activists as well as the scientists such as Stephen Schneider and James Hansen, who have invested major effort in the study of climate change.

The governmental actions that are being considered could have profound effects on the economies of many nations, and for that reason, many economists, as well as many scientists, urge caution and more study before taking drastic action. It is important that policymakers consider how much effect limiting or reducing carbon dioxide emissions will have on the atmosphere and at what economic and social costs.

A BRIEF HISTORY OF GREENHOUSE POLITICS

Before discussing the various proposed actions and their potential consequences, let's see how government became involved in the global warming debate.

Although the greenhouse effect was described in the 1820s, and even associated with carbon dioxide in the 1860s, it was well over a century before the issue was noticed outside the scientific community.

During the 1960s an increase in environmental research, aided by the development of mainframe computers, stirred up some political interest. Conferences were held and reports were written, but no recommendations for action were made, and other, more compelling issues held the public's attention, issues ranging from Vietnam to the race for space, from the population explosion and civil rights to "sex, drugs, and rock and roll."

In the seventies, disillusioned with big business and buoyed by campus demonstrations, the spirit of Woodstock, and a liberated lifestyle, the "new generation" swept the country; the time was ripe for awareness and activism. Environmental issues became immensely popular, and global cooling caught the imagination of hippies and politicians alike—it was the decade of one of the coldest winters on record. It was also during this decade that scientists became worried about deterioration of the ozone layer, a concern that led to the ban of chlorofluorocarbons (CFCs) in aerosol cans. (The CFCs were replaced with propane and butane, highly volatile gases that pose serious fire safety hazards.)

The presidential administrations of the seventies, primarily those of Nixon and Carter, were committed to environmental action. Under Richard Nixon, the Clean Air Act of 1970 was enacted, giving extensive power to the federal government. Earlier attempts to cope with dramatic increases in air pollution had left most of the jurisdiction to the states, with the federal government acting as technical advisor and financial assistant. With little progress being made, Congress stepped in with the Clean Air Act, which created a whole set of standards and compliance requirements. The U.S. Environmental Protection Agency (EPA) was created at this time and given the authority to set the standards.[2] It remains the most powerful environmental agency in the country and perhaps even in the world.

Under Jimmy Carter there was extensive support of alternative energy, paving the way for major research and development on renewable energy technologies, especially solar energy. The demand for energy had grown rapidly in the 1960s and early 1970s. Energy was inexpensive and little thought was given to efficiency. However, according to a 1993 *American Scientist* article, the "oil shocks" of the early and late 1970s, when OPEC (the Organization of Petroleum Exporting Countries) raised oil prices, led to concern and changes in the way Americans used energy. Public concern was high in 1977 when Carter declared that energy use was the most important issue facing the country. Fuel conservation and efficiency became popular, leading to a further slowdown in the demand for energy.[3] Ironically, this slowdown alleviated much of the concern over air pollution and lack of fuel supplies, and it set the stage for the 1980s, when energy prices would drop once again and environmental problems would take a back seat to more pressing issues.

The decade of the seventies ended with the First World Climate Conference in Geneva, Switzerland. By this time the climate predictions had shifted from cool to warm, and this 1979 conference covered environmental issues such as the greenhouse effect and climate change, prompting further research and a heightened interest among scientists and governmental agencies.

The following year, the United Nations Environment Programme (UNEP), the World Meteorological Organization (WMO), and the International Council of Scientific Unions (ICSU) sponsored a meeting in Villach, Austria. The attendees concluded that carbon dioxide-induced climate change was a major environmental issue and agreed that more research was needed for a firm scientific base. These organizations pointed out the need for global cooperation between developing and industrialized nations.

In that same year, Ronald Reagan was elected president of the United States, and there was a decided shift in environmental philosophy in the government. Reagan came under fire from environmentalists because he was a friend of business and industry, and, in fact, it was under the Reagan administration that the budget for the EPA was slashed dramatically and federal investment in solar energy and other renewable energy sources virtually ended. In addition, the demand for energy had dropped, and energy prices as well, so that the public political climate favored Reagan's position.

In 1983 the EPA issued a report warning of a 2°C rise in global temperature and determined that the only way to avoid it was to ban the use of coal before the year 2000.[4] But at the same time the National Academy of Sciences (NAS) issued a report downplaying the problem, saying that they did "not believe . . . that the evidence at hand about CO_2-induced climate change would support steps to change current fuel-use patterns away from fossil fuels."[5] In addition, the NAS report recommended

> caution in undertaking any major changes in current behavior and policies solely on account of CO_2. It is probably wiser not to act aggressively . . . right now when we really do not know the future consequences or context of CO_2 increase.[6]

In spite of this inconsistency between the positions of the EPA and the NAS, the idea of an environmental calamity quickly caught the imagination of the news media.

In 1985, a second conference was held in Villach, Austria, sponsored once again by UNEP, WMO, and ICSU, and attended by research scientists, ecologists, and computer modelers. According to Sonja Boehmer-Christiansen's 1994 commentary in *Nature*, this symposium attracted support from major environment/energy research bodies but was attended by only two government scientists, one from the United Kingdom and one from the United States (Department of Energy). Thus the primary interests being served were the contract research institutions with interests in carbon dioxide and climate variability. Boehmer-Christiansen points out that, with such a strong warning about the dangers of global warming, "stringent regulations would be needed . . . opening up energy markets to 'green' technologies."[7] The warnings spread, and the political momentum needed for major governmental action was rapidly building.

Scientific meetings, press coverage, confirmation of the ozone "hole," and a fully engaged environmental movement resulted in the 1987 Montreal Protocol, the first international treaty based on the results of computer models! The main purpose of the treaty was to eliminate CFCs entirely because of their implication in the diminishing stratospheric ozone. Shortly thereafter, the greenhouse effect of CFCs was recognized, and the Montreal Protocol was amended to hasten the phase-out period of these chemicals.

The year 1988 stands out as a watershed year in the story of global warming policy. That year's extremely hot summer, along with droughts in the United States and flooding in Bangladesh, provided the ideal setting for the global warming debate to shift dramatically to the side of the alarmists.

In June an international conference in Toronto, "The Changing Atmosphere: Implications for Global Security," called for reduction of carbon dioxide emissions by approximately 20% of 1988 levels by the year 2005. The conference proposed a world fund, supported by a carbon tax, to help developing nations cope with the special economic problems they faced in complying with emissions reductions. In addition, the Intergovernmental Panel on Climate Change (IPCC) was created by WMO and UNEP to coordinate worldwide climate research efforts. This organization is now the foremost international authority on climate change. The reports published by the panel are used as a major source of information for policymaking. (In addition to the IPCC, NAS, as official scientific advisor to the United States government, has substantial influence on federal policy, and its policy recommendations provide the framework for national action plans and policy decisions.)

In 1988 George Bush was elected president after campaigning as the "environmental candidate." However, after Bush took office, he continued Reagan's cautious approach to environmental policy and found himself criticized by environmental organizations and by other countries.

Perhaps the most significant global warming event of 1988 was Jim Hansen's testimony before Congress that greenhouse warming had arrived. While infuriating many of his colleagues, as we discussed in Chapter 1, Hansen nevertheless opened up the discussion politically, paving the way for funding of additional climate research.

Hansen's testimony was controversial in more ways than one. When the Office of Management and Budget (OMB), which monitors federal policy statements, insisted the following year on attaching a qualifying statement to his written testimony, cries of censorship were heard. Then-Senator Al Gore considered the OMB's act to be part of Bush's effort to downplay environmental issues: "Why would the Bush White House go to such lengths to avoid facing the facts about the environment? Is it because the necessary changes would . . . cause some political risk?"[8] And what was the OMB's caveat that caused such a commotion? This is the statement that was added: ". . . these changes should be viewed as estimates from evolving computer models and not as reliable predictions."[9] In fact, this statement is consistent with the qualifying statements that appear in essentially all scientific reports on climate change research.

The international call for action became louder as the European Economic Community (EEC) criticized America for inaction, and in 1989 the European Council of Environment Ministers called for an immediate response to the global warming crisis regardless of uncertainties. The official White House position, which was to wait and see before taking actions that might be inappropriate further down the road, was influenced by a report from the George C. Marshall Institute, according to David E. Newton of the University of San Francisco, in his reference handbook *Global Warming*.[10] The Marshall Institute provides scientific advice for public policymakers, and its conclusions, criticized by the global warming alarmists, support a conservative view of potential climate change. Ironically, at the very time the United States was being criticized for inaction, Congress was passing its amended Clean Air Act, which would cost $20 billion per year.

At international conferences in 1989 and 1990, President Bush blocked proposals for specific limitations of carbon dioxide emissions because of his concern for the economic impacts of drastic reductions. His actions upset environmentalists worldwide and led to further criticism of his administration.

In 1990 IPCC published its first report, *Climate Change*, which supported the view that emissions resulting from human activities were increasing atmospheric greenhouse gas concentrations and would contribute to a warming of the earth's surface. While stopping short of making actual recommendations for action, the report stated that an immediate reduction of 60% in anthropogenic emissions would be required to stabilize the concentrations at 1990 levels.[11] The ambiguity of the report, however, led to a wide range of interpretations. The global warming adherents pointed to the warnings in the report while detractors quoted the inconsistencies.

As heat records continued to pile up and *Nature* reported a decrease in the Arctic ice cap, the international debate intensified. The 1992 United Nations Conference on Environment and Development (UNCED) in Rio de Janeiro, a conference better known as the Earth Summit, called for drastic reductions in carbon dioxide emissions. President Bush further alienated himself from environmental activists when he announced that he would not sign the treaty unless specifics about the levels at which carbon dioxide would be stabilized were eliminated. Concessions were made, leaving out specific emissions levels and time frames, but the United States was criticized for watering down the agreement. (A political cartoon that year showed Bush introducing himself as the Environmental President to a panel at the Earth Summit. The panel members in turn introduced themselves as the Easter Bunny, Joan of Arc, and the Tooth Fairy.)

The Framework Convention for Climate Change that was agreed upon at Rio was formulated by the IPCC. The prime goal of the agreement was to stabilize atmospheric greenhouse gases in order to prevent "dangerous" anthropogenic interference with the climate system. However, according to Bert Bolin, IPCC's chairman, the agreement contained language and concepts that should be clarified, such as the term "dangerous."[12] So while recommending action, no specifics were given, leading to a great deal of flexibility in interpreting the responsibilities of the various governments.

In 1993 a new "environmental president" took office, along with an "environmental vice president." Although Vice President Al Gore still takes an extreme view of the dangers of global warming, as evidenced by his pronouncement on Earth Day 1994 (Chapter 2), the Clinton administration is more cautious than environmental activists would like. These activists have criticized the administration's Climate Action Plan of 1993 for favoring industry. The plan relies mostly on voluntary programs and increased use of renewable energy sources, and it calls for "limited, and focused, government action and innovative public/private partnerships."[13] A spokesperson for Greenpeace, one of the more aggressive environmental organizations, complained that "all we're getting is 'greenwash' without any real change in energy policy," while another one said that the "very industries called upon to be partners in the plan believe climate change is a myth."[14] While the Clinton plan continues to be lambasted by the activists, it is buying time for further research and consideration of the consequences of carbon dioxide emissions reductions.

CHARTING TODAY'S COURSE

With so many organizations—some scientific, some environmental, some political—involved in trying to influence policy responses, you can be sure of a wide variety of recommendations. The IPCC's scientific assessment is used as a basis for much of the policy being made, but, as pointed out in the 1990 report, it is a summary of scientific understanding at the time of the report, not a definitive statement about climate change.

> Uncertainties attach to almost every aspect of the issue, yet policymakers are looking for clear guidance from scientists; *hence authors have been asked to provide their best-estimates wherever possible* [IPCC emphasis], together with an assessment of the uncertainties.[15]

The IPCC members clearly recognize the perilous position they are in. They have been criticized both for being too conservative and for overreacting.

The 1992 IPCC report includes some updates that would affect much of the policy response being considered. Of particular note is its reevaluation of Global Warming Potential (GWP), a rating that represents the relative potential climate effect of greenhouse gases. The GWP is used as a tool in determining which gases should be considered when calling for emissions reductions. (It was used to rate CFC replacements, and some, which were otherwise acceptable, were eliminated because of the GWP, leaving very few products that can be used as substitutes.) The 1992 IPCC report reveals increased uncertainty in the GWP calculations.[16] If these calculations cannot be trusted by scientists, how can they be used as a basis for determining the amount of reduction needed in carbon dioxide emissions? If scientists cannot be sure of the effects of carbon dioxide emissions, how can they determine the effects of reductions?

Most discussions of global warming focus on anthropogenic carbon dioxide, which is primarily the result of fossil fuel burning and deforestation. (Keep in mind that the amount of atmospheric carbon dioxide caused by man's activities represents about .007% of the chemical composition of the atmosphere—see Chapter 6.) Since the industrial revolution, these factors have led to an increase of about 26% in carbon dioxide concentration in the atmosphere, according to the IPCC.[17] Population and economic growth will greatly influence future levels of emissions.

Most of today's policy discussion revolves around the following categories of mitigating the effects of greenhouse gases or adapting to climate change:

- Options that eliminate or reduce greenhouse gas emissions in order to slow or prevent warming.
- Options that offset emissions by removing gases from the atmosphere, blocking incoming solar radiation, or altering reflection of radiation.
- Options that help human society and ecosystems adapt to climate change.

Within each of these categories there are a number of possible actions with varying consequences. Some of them are very expensive, while others have potentially extreme adverse effects. Let us look at the actions being considered.

REDUCING GREENHOUSE GAS EMISSIONS

President Clinton's 1993 Climate Change Action Plan calls for returning U.S. greenhouse gas emissions to 1990 levels by the year 2000, in accordance with the Rio treaty, which requires a 20% reduction in emissions. What effect would this level of reduction have? In 1990 the IPCC calculated that an immediate reduction of over 60% of anthropogenic carbon dioxide, nitrous oxide, and CFCs would be required to stabilize atmospheric concentrations, along with a 15–20% reduction in methane. Clearly, a 20% reduction in overall emissions would not meet the goal of stabilizing greenhouse gas concentrations.

There are various actions that can be taken to reduce greenhouse gas emissions. The methods most commonly considered include the following:

Improving Energy Efficiency and Conservation

This is no doubt the single most effective strategy, and it has additional benefits:

- Reduction of air pollution.
- Increased energy independence. (Remember the Persian Gulf War?)
- Additional time to study potential climate effects and allow for advancement of technology.
- Competitive advantage of lowered cost of manufactured products. (The United States has the highest per capita cost of energy, and this high cost is reflected in the costs of manufacturing.)

Developing Alternate Energy Sources

If reliance on carbon-containing fuels can be reduced by the development of energy sources that do not emit greenhouse gases, overall emissions should be reduced. The alternate energy sources mentioned below have all been discussed and developed to some degree.

Solar. Solar energy is one of the most promising alternative energy sources. Scientists have developed photovoltaic cells that are close to being cost competitive with fossil fuel power plants. There are currently isolated areas in which solar energy is in fact less expensive.

Wind. Windmill production enjoyed a surge during the energy crises of the 1970s. Development continued during the eighties, with smaller and smaller machines being built. According to Eliot Marshall, writing in *Science*, the wind energy business is booming in the nineties thanks in part to laws that encourage the use of wind-generated electricity. In 1993, there were more than 16,000 wind turbines in the United States, most of them in California.[18] Of course, windmills require huge land areas located in areas of constant or near-constant winds.

Geothermal. Most experiments with geothermal energy sources have ended in failure. There are few reliable sources of hot springs, and many of them are very corrosive because of their mineral content. The occasional steam vents have also proven to be corrosive. There is little hope that this energy source will be exploited with a perceptible impact on the world's energy requirements.

Biomass. The development of biomass fuel sources will continue to be a slow process. Biomass fuels are all carbon-based and will have emissions of carbon dioxide associated with them. It is probable that the supply would not keep up with the demand, even if fast-growing forests are utilized for energy requirements. The costs of fossil fuels will have to be extreme before biomass will provide more than the occasional demonstration experiment.

Nuclear. Nuclear power can be and is a significant alternate energy source in many countries, including the United States. Currently, about 21% of the electric power in the United States is generated by nuclear energy; however, no new plants have been ordered since the late 1970s. The fear of nuclear accidents and the issue of radioactive wastes have made the introduction of new reactors too risky for the power utilities. In addition, the costs of building a nuclear power plant have skyrocketed largely due to government regulations and poor construction management by the utilities. When the costs of fossil fuels reach a high enough level, the world will once again turn to nuclear power in spite of its problems.

Removing Harmful Gases (Scrubbing) before They Reach the Atmosphere

The problem here, of course, is what to do with the carbon dioxide. One solution being studied is to inject it into abandoned oil or gas wells, but the cost is exorbitant and would probably raise the cost of electricity 30% or more, according to Theodore Simpson, of the United States Department of Energy.[19]

Another solution calls for injecting CO_2 into the ocean. The earth's oceans hold many times the amount of carbon dioxide present in the atmosphere and are capable of holding much more. Unfortunately, the process of removing it from exhaust gases, compressing it, and piping it into the ocean appears to be quite expensive. In addition, there could be unwanted side effects. The carbon dioxide-rich water would be acidic and possibly deadly to organisms, and the dissolved carbon dioxide could cause seawater to lose much of its dissolved oxygen, another life-threatening effect.

Ironically, federal funding for research and development in the areas of renewable energy, conservation, and nuclear fission has decreased. According to Rosina Bierbaum and Robert M. Friedman of the Office of Technology Assessment, the combined budget for these three areas was 80% lower in 1990 than it was in 1980, and bringing it up to 1980 levels would cost $2.6 billion. This expenditure would hasten the development of technologies that would reduce greenhouse gas emissions.[20] This seems like a small price to pay for an investment that would have many benefits.

OFFSETTING GREENHOUSE GASES

Offsetting greenhouse gases after they have entered the atmosphere may be accomplished by increasing the natural carbon sink capacity of forests and by technological fixes.

Forests

Methods of modifying the carbon cycle and thereby offsetting increased carbon dioxide in the atmosphere are somewhat poorly understood. The capacity of forests to offset emissions is limited; once a tree dies or is cut down, it releases its stored carbon back into the atmosphere when it decays or is used as fuel. Even so, the deleterious effects of deforestation are generally recognized and options have been suggested for dealing with the forestry issue:

- Slow or stop the loss of existing forests. Unfortunately, large areas of land that are subject to deforestation are in tropical countries where competition for available land is intense and governmental restrictions are inadequate.

- Accelerate reforestation. George Woodwell, director of the Woods Hole Research Center in Massachusetts, calculated that two million square kilometers, or about 770,000 square miles, of new forest would be needed to remove one billion tons of carbon dioxide annually. (About three billion tons would need to be removed each year to stabilize atmospheric carbon dioxide concentrations.)[21] Two million square kilometers represents an area the size of Texas, California, Arizona, New Mexico, and Nevada combined, or a fifth of the United States!
- Adapt forest management practices to increase the carbon stored in nonliving reservoirs, such as agricultural soils, and in artificial reservoirs, including timber products.

Geoengineering

Projects that will offset climate change through technology have been suggested for a long time. You may recall the discussion in Chapter 2 about geoengineering plans to protect the planet from global cooling in the seventies followed by recent suggestions for offsetting warming potential. Recognition that the climate system is still poorly understood demands caution in applying these options, considering their potentially adverse effects on climate and other side effects, as well as the economic costs.

Screening Incoming Radiation. One such proposal is to shoot smart mirrors into space with rifles in order to reflect sunlight, screening it from the earth's atmosphere. Another involves putting dust or soot in orbit for the same purpose, as in a project proposed by Wallace Broecker, a geochemist at Columbia University. He suggests deploying a fleet of aircraft to load the stratosphere with sulfur dioxide. In order to offset the predicted doubling of carbon dioxide concentrations, thirty-five million tons of sulfur dioxide would need to be added per year. This could be accomplished by a fleet of 700 jumbo jets working around the clock every day, at an annual cost of about $20 billion.[22] Those scientists who believe that we should be worrying about global cooling caused by atmospheric particulates (Chapter 9) would be quick to point out the folly of such a project.

Stimulating Cloud Formation. This project would change cloud abundance by increasing cloud condensation nuclei through carefully controlled emissions of particulate matter. The resulting clouds would both reflect and absorb solar radiation, cooling the earth. However, stratospheric particles are implicated in the depletion of the ozone layer.

Seeding the Ocean with Iron. Dubbed the Geritol solution by some, this plan involves dosing millions of square miles of ocean with iron dust, which would fertilize phytoplankton (microscopic plants). The phytoplankton, in turn, would absorb carbon dioxide, then die and sink to the bottom of the sea, storing away the carbon. While this idea worked well in theory and in small-scale laboratory tests, the practical application has been disappointing. In experiments carried out in the Pacific by marine biologists from Duke University, the iron supplement quickly sank to the bottom of the ocean, leaving the surface waters as iron deficient as they already were.[23] Another problem with this approach is the biological response to an increase in phytoplankton, according to an article in *Chemical & Engineering News,* which reported on research into iron supplementation. It turns out that an increase in phytoplankton ultimately results in an increase in the numbers of grazing zooplankton, microscopic marine animals. Because zooplankton breathe out carbon dioxide, they counteract any positive benefits of the absorption of CO_2 by phytoplankton.[24]

Geoengineering projects must be pursued with caution. Deliberately tampering with nature can have far greater consequences than the accidental effects so far created by society.

ADAPTING TO CLIMATE CHANGE

The simplest, and cheapest, approach to climate change is to learn to adapt to it. Natural climate variability has always existed and will continue to do so, regardless of whether or not there are anthropogenic causes. Moreover, large numbers of people live in virtually all the earth's climate zones and move about between them.

Assuming the worst effects of warming, some of the areas most affected would be agriculture, water systems, and preservation of biodiversity. The NAS study of 1991 included the following recommendations for adaptation:

Agricultural research. Agriculture is the activity most susceptible to climate change. As climate variability has always been a fact of life, agricultural practices have had to change. Special areas that researchers should explore are:

- Sustaining natural resources
- Remaining productive during extreme weather conditions
- Becoming more water efficient
- Exploiting the fertilization benefits of increased CO_2 (a point usually overlooked by global warming apostles)

Water management. Weather and precipitation cause natural variability in the water supply. Coping with this present variability can help to prepare for future climate change by increasing efficiency of water use through rights, markets, and prices, and by better management of present systems of supply.

Biodiversity. One of the factors most commonly emphasized as a consequence of global warming is an increase in the rate of loss of biodiversity. Actions recommended include:

- Establishing and protecting habitats
- Conducting an inventory of little-known species
- Collecting key organisms

- Controlling and managing wild species[25]

As with most of the global warming issues, adapting to climate change has its advocates and its detractors. There are those who believe that the concept of adaptation will get in the way of needed action, that if we believe that adaptation to climate change is possible, the will to act politically will be diminished. Al Gore, in his book *Earth in the Balance,* warns that faith in our ability to adapt is a sort of "laziness in our spirit," that we have been "seduced by industrial civilization's promise to make our lives comfortable."[26] And yet industrial civilization has done far more than make our lives comfortable. It has contributed to a healthier populace with longer life spans and more efficient use of natural resources.

Many scientists believe climate change will be so gradual that there will be ample time to adjust. One factor that is usually overlooked in the doom-and-gloom approach to making policy is the rapid advance of technology. Current predictions of temperature increase indicate that any changes will take place gradually, allowing time for adapting and for developing countermeasures, if needed. William Nierenberg, Director Emeritus of the Scripps Institution of Oceanography, points out the folly that would have resulted if the world had reacted seriously to the predictions of a 25-foot sea level rise, making unneeded engineering investments at great cost and little value.[27] These predictions of sea level rise have steadily decreased with additional research; the current predictions are measured in inches rather than feet. In another fifteen years, will we be warned of decreasing sea levels?

HOW HIGH THE COSTS?

Climate change has potentially enormous economic impacts. There are costs if we do nothing and warming occurs; there are costs of taking drastic action.

Costs of the Effects of CO_2 Doubling

What will be the economic effect of global warming, if it occurs? Let's look at some varying opinions of the effect on the United States.

Global warming adherents usually point to agriculture as the primary victim of climate change and warn of food shortages. Water supplies would also be affected, both for irrigation and hydroelectric power generation. Since most of the effects are predicted in terms of social costs and limited lifestyles, they are not easily quantified in dollar amounts.

However, some economic analysts have tried to put these predictions into numbers. William D. Nordhaus, an economist from Yale University, is a well-known authority on the economics of global warming. At a global warming conference in 1990, Nordhaus discussed the impact of climate change on human society and the relative economic effects of various policy responses. According to Nordhaus, the advanced countries, such as the United States, Canada, Japan, and Western Europe, have developed to the point where climate variables like temperature and humidity have little effect on economic activities, and climate is no longer much of a consideration in locating businesses. Of far greater importance are human factors, such as wages, unionization, labor-force skills, and political climate. So while the temperature variable, which is the primary focus of computer climate models, is useful as an index of climate change, the important climate variables are precipitation or water levels, extremes of droughts or freezes, and the size of projected changes as compared to day-to-day changes.

In Nordhaus's discussion of the economic effects on the United States of a doubling of carbon dioxide (the basis used for most of the computer climate models) and a global temperature increase of 3°C, he concludes that the effects would be small—around one-quarter of 1% of national income.[28] In a subsequent article, Nordhaus estimates that this would be about $15 billion at 1992 prices.[29] Considering that the population of the United States was 249 million in 1990, the per capita cost of the effects of such global warming would be about $60 per year. Keep in mind that the IPCC's estimate of the expected warming lies in the range of 1.5–4.5°C.

Nordhaus goes on to state that, while there is not as much statistical information available for other countries, he estimates that there would be little effect in other advanced countries. However, small countries that are heavily dependent on coastal activities, or that suffer major climate change, would be most affected. Since most poor countries rely on agriculture, the positive effects of carbon dioxide enrichment might offset damages. Those living on "the ragged edge of subsistence with few resources to divert to dealing with climate change" would be most adversely affected.

Nordhaus's conclusion is that climate change will produce a combination of gains and losses that will not be noticeable compared to other changes, and says "those who paint a bleak picture of desert Earth devoid of fruitful economic activity may be exaggerating the injuries and neglecting the benefits of climate change."[30] Considering the devastating impact of such uncontrollable events as earthquakes and volcanoes, the economic effects of global warming start to look somewhat negligible.

Costs of Emissions Reductions

In 1989 the EPA calculated that stabilizing U.S. carbon dioxide emissions would force 30% taxes on oil and coal, whereas meeting the demands for a 20% reduction in emissions, the goal of the Climate Action Plan as well as the Rio treaty, would require a tax of $25 per barrel on oil and $200 a ton on coal.[31] This would double U.S. energy costs.

Remember that the Department of Energy (DOE) figured the cost of a 20% reduction in carbon dioxide emissions to be as much as $95 billion per year.[32] And according to the Office of Technology Assessment, under a "tough" scenario of reductions, which would lower emissions as much as 35% (over 1987 levels) by 2015, the United Sates could save as much as $20 billion a year or lose as much as $150 billion, based on the costs versus the fuel savings.[33] This wide range of estimates reflects the many variables involved in making economic projections and serves as a reminder that such projections share the uncertainties exhibited by computer climate predictions. Consider the aforementioned estimates of the economic effects of global warming. A price tag of $15 billion for business as usual, as estimated by economist Nordhaus, seems small compared to a potential cost of $150 billion for lowering emissions.

Clearly, Third World countries, who are increasing their emissions at a far greater rate than the developed countries, are unable to meet such demands. The 1992 Rio treaty did not include requirements for developing countries to curtail emissions. The issue of developing versus developed countries includes more than just relative effects of emissions on the atmosphere, as we shall discuss shortly.

Carbon Tax

A common solution that is frequently bandied about is the carbon tax. Georgii S. Golitsyn, of the Moscow Physical Technical Institute, put it this way at the Greenhouse Glasnost, a conference organized by Robert Redford through his Institute for Resource Management: "I propose taxing every ton of fossil fuel consumed in developed nations at a small percentage of its cost. From these proceeds an international assistance fund could be established for developing nations to introduce energy- and resource-saving technologies . . ."[34] Sounds good, but carbon taxation is a very complicated issue. How it is to be applied internationally, whether or not the taxes are applied as production or consumption taxes, who should get the revenues, and how the carbon content is calculated are all questions that contribute to a problem requiring a great deal of study. Nevertheless, taxing coal-based fuel usage could have the effect of encouraging a shift to renewable fuels as well as energy efficiency and conservation.

POLITICAL SENSITIVITY

The most common response to the position that burning of fossil fuels contributes to global warming is to call for cutbacks in their use (and in fact that approach is included in the 1993 Climate Change Action Plan, which calls for returning U.S. greenhouse gas emissions to 1990 levels by the year 2000.[35]) But the buildup of carbon dioxide is a global phenomenon because of atmospheric circulation, and if emissions are to be controlled, the effort will have to be made on a worldwide basis.

And although it is a global problem, it is not an equitable one. All nations are not equally responsible. In fact, as industrialized countries have economized on labor and capital by shifting away from coal and toward oil, gas, and carbon-free fuels, their per capita carbon dioxide emissions have been reduced, while developing countries have increased their carbon dioxide emissions. Figures 3.1 and 3.2 show the carbon dioxide emissions and the per capita emissions for selected industrialized and developing countries. Note that while the carbon dioxide emissions of the developed nations have leveled off or even decreased, the developing nations of India and China are increasing both their total and their per capita emissions. The developing nations are also increasing their populations at much faster rates than the developed nations. These trends should be expected for all the developing countries.

All nations are not able to cut emissions equally, and the developing countries would need a great deal of assistance to do so and still continue to develop. All nations are not equally affected by climate change. Some would actually benefit from increased rainfall and a more temperate climate, while others would suffer from coastal erosion and flooding.

Fred Singer, well-known climatologist, pointed out the realities of these inequities in 1989, when he wrote that because fossil fuel is a major economic commodity, with tremendous investments already incurred by the majority of countries, it would be economically disastrous to abandon it as an energy resource. Since carbon dioxide buildup is a global phenomenon, any course of action would have to be agreed upon and implemented internationally. But because the perceived climate change will actually be advantageous for some areas, those areas might be reluctant to make sacrifices.[36] Since the scientific community is still in disagreement

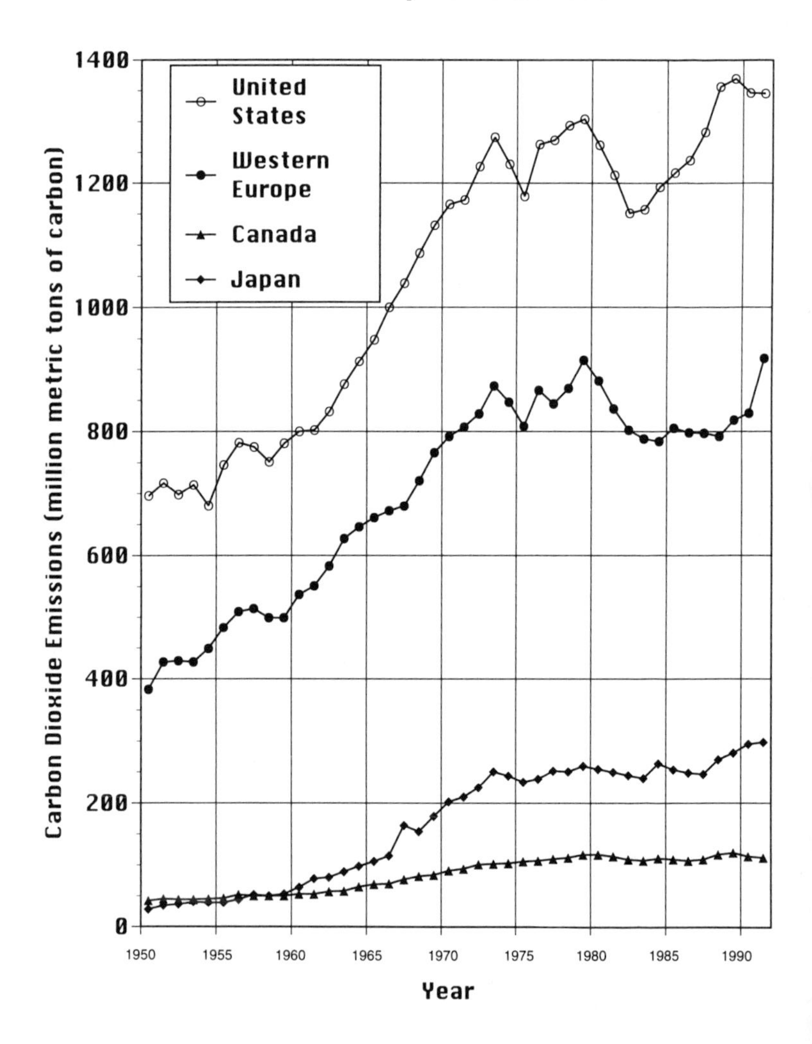

FIGURE 3.1. Carbon dioxide emissions expressed as millions of metric tons of carbon from 1950 to 1991: a, developed countries; b, developing countries. The data for these figures were taken from G. Marland, R. J. Andres, and T. A. Boden, "Global,

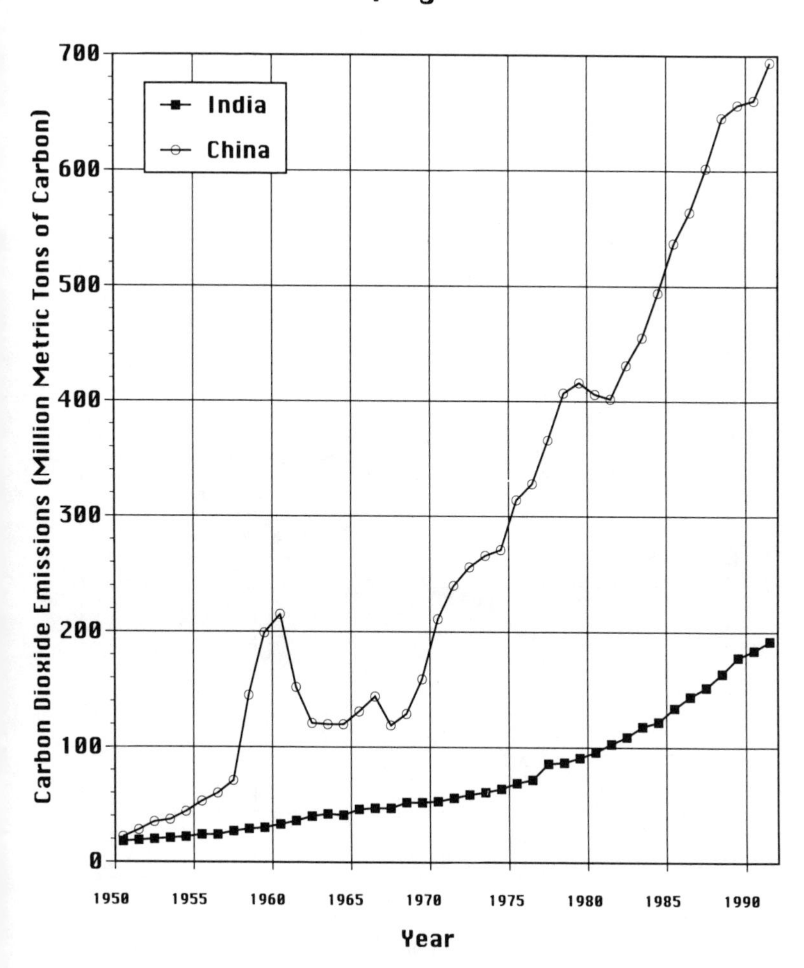

Regional, and National CO_2 Emissions," in *Trends 93: A Compendium of Data on Global Change*, ORNL/CDIAC-65, ed. T. A. Boden, D. P. Kaiser, R. J. Sepanski, and F. W. Stoss (Oak Ridge, TN: Oak Ridge National Laboratory, 1994), pp. 505–584.

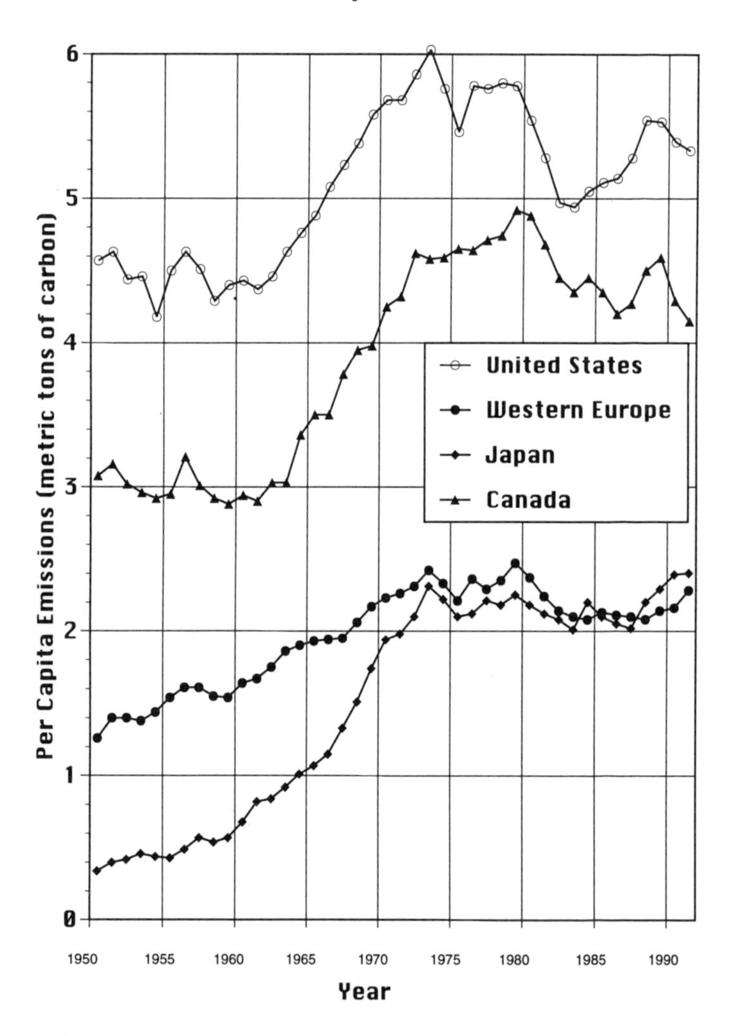

FIGURE 3.2. Carbon dioxide emissions per capita expressed as metric tons of carbon per person from 1950 to 1990: a, developed countries; b, developing countries. The data for these figures were taken from G. Marland, R. J. Andres, and T. A.

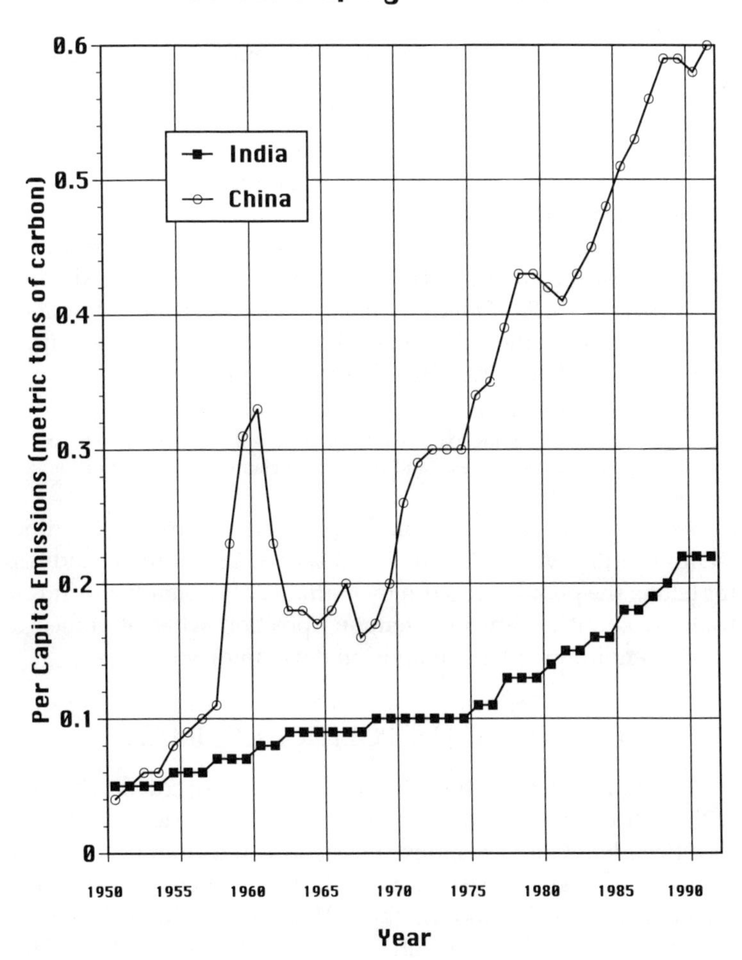

Boden, "Global, Regional, and National CO_2 Emissions," in *Trends 93: A Compendium of Data on Global Change*, ORNL/CDIAC-65, ed. T. A. Boden, D. P. Kaiser, R. J. Sepanski, and F. W. Stoss (Oak Ridge, TN: Oak Ridge National Laboratory, 1994), pp. 505–584.

about the future climate, many people are reluctant, and rightly so, to take drastic action that would have a profound effect on the economy.

There is also the sensitive issue of resentment, on the part of developing countries, of the developed countries who have had the advantages of an increased standard of living because of the very activities that created the problem, and who wish to deprive developing countries of the same benefits. As pointed out in a 1989 *Forbes* article, "there is no way to prevent a [Third World] CO_2 doubling without slashing growth and risking a revolt of the have-not nations against the haves."[37] This reality was reinforced at Rio. In a 1992 article in the *San Francisco Chronicle*, Maurice Strong described criticism hurled at the industrialized countries by a negotiator from Malaysia, Wen Lian Ting. Wen insisted that unless

> rich countries pay Malaysia not to cut its trees . . . the cutting will continue. . . . Malaysia is not going to keep its trees "in custody for those who have destroyed their own forests and now try to claim ours as part of the heritage of mankind!"[38]

An editorial in *Nature* reminds us: "Rich countries in the nineteenth century were called imperialists because of their fondness for telling the poor how to run their affairs."[39] Certainly the industrialized countries are in a tenuous position when it comes to making environmental demands on developing countries.

CHANGING U.S. POLITICAL CLIMATE

On a national level, the political climate has, of course, changed with each presidential administration. We've already mentioned Vice President Gore's response to scientific uncertainties. While a United States Senator, Gore was on the Committee on Commerce, Science and Transportation. In the 1991 hearing on the NAS's report "Policy Implications of Greenhouse Warming," Gore criticized the report (and the scientists) for being overly cautious. Frank Press, president of NAS, responded:

> Sometimes it is not a question, it seems to me, of being cautious; it is a question of what makes good sense based on scientific evidence.

> Sometimes we get ahead of our science. I suspect the Soviets thought when they built Chernobyl that they were in full grasp of all of the science necessary to build and effectively run that plant. They were not. So, you can sometimes go too fast, and too far, just as you can sometimes be too cautious and too slow.
>
> It is not an easy task to pick the right path . . .[40]

Press's comments serve as a chilling reminder of the events of Chernobyl and their effect on the fears of the use of nuclear energy. The lesson, of course, is not that nuclear energy is to be feared, but that proceeding too fast in the face of uncertainties is risky.

The IPCC also finds itself in a politically sensitive situation because of its influence on policymaking. Some scientific journals and the popular press have criticized IPCC conclusions "for lack of openness about uncertainties and for brushing aside controversies."[41] While we have found the reports to mention the uncertainties, we have found the statements couched in qualifiers. And it is somewhat difficult to reconcile the fact that IPCC states the uncertainties while still drawing the conclusions that appear to be politically correct, conclusions that recommend taking drastic action.

FLYING TOO CLOSE TO THE SUN

Remember Daedalus and Icarus. The hysteria of the global warming myth threatens to be the fuel that will fly us too high, too soon. In 1984, Bernard Cohen, a physicist at the University of Pittsburgh, warned of the dangers of scientific policy being influenced by the emotional responses of the public when he said:

> Our government's science and technology policy is now guided by uninformed and emotion-driven public opinion rather than by sound scientific advice. Unless solutions can be found to this problem, the U.S. will enter the 21st century declining in wealth, power and influence. . . . The coming debacle is not due to the problems the environmentalists describe, but to the policies they advocate.[42]

Cohen was rightly concerned. As we have seen, public policy is being increasingly influenced by the loud cries of environmental activists and the scientists who strongly support the global warming theory.

The warnings continue. A 1994 editorial in *Science* warns of the power of environmental organizations:

> Concern continues that the U.S. Environmental Protection Agency (EPA) might curtail or ban the production of chlorine and compounds containing it. This perception has been fostered by indications that EPA policy is being predominantly influenced by Greenpeace and its allies.[43]

Remember that the EPA was established as an agent of the federal government to set air quality standards and demand compliance, and it is possibly the most powerful governmental agency on earth. Careful consideration must be given to any policy that will give it more power, as would happen with more stringent emission standards.

As we shall see in subsequent chapters, the foundation on which the global warming theory is built is full of inconsistencies and inadequacies. It is imperative that the uncertainties in the science of the global warming theory be cleared up before irreversible policy decisions are made.

REFERENCES

1. Committee on Science, Engineering, and Public Policy (COSEPUP), *Policy Implications of Global Warming* (Washington, DC: National Academy Press, 1992), p. 67.
2. Roger W. Findley, *Environmental Law in a Nutshell* (St. Paul, MN: West Publishing, 1992), pp. 99–100.
3. John Ahearne, "The Future of Nuclear Power," *American Scientist* January-February 1993:24–25.
4. Annika Nilsson, *Greenhouse Earth* (Chichester: John Wiley & Sons, 1992), p. 8.
5. Carbon Dioxide Assessment Committee, *Changing Climate* (Washington, DC: National Academy Press, 1983), p. 4.
6. Ibid., 62.
7. Sonja A. Boehmer-Christiansen, "A Scientific Agenda for Climate Policy?" *Nature* 372 (1994):401.
8. Al Gore, *Earth in the Balance* (Boston: Houghton Mifflin, 1991), p. 176.
9. Richard A. Kerr, "Hansen vs. the World on the Greenhouse Threat," *Science* 244 (1989):1041.
10. David E. Newton, *Global Warming* (Santa Barbara, CA: ABC-CLIO, 1993), p. 36.
11. J. T. Houghton, G. J. Jenkins, and J. J. Ephraums, eds., *Climate Change: The IPCC Scientific Assessment* (Cambridge: Cambridge University Press, 1990), p. xi.

12. Bert Bolin, "Next Step for Climate-Change Analysis," *Nature* 368 (1994):94.
13. President William J. Clinton and Vice President Albert Gore, Jr., *The Climate Change Action Plan* (Washington, DC: United States Government, 1993), p. 1.
14. Pamela Zurer, "Environmental Groups Fault Climate-Change Plan," *Chemical & Engineering News* 9 May 1994:28–29.
15. Houghton, Jenkins, and Ephraums, xiii.
16. J. T. Houghton, B. A. Callander, and S. K. Varney, eds., *Climate Change 1992* (Cambridge: Cambridge University Press, 1992), p. 5.
17. Houghton, Jenkins, and Ephraums, xv.
18. Eliot Marshall, "A Fair Wind Blows for One Green Technology," *Science* 260 (1993):1887.
19. Theodore B. Simpson, "Limiting Emissions of the Greenhouse Gas, CO_2," *Environmental Progress* 10 (1991):248.
20. Rosina Bierbaum and Robert M. Friedman, "The Road to Reduced Carbon Emissions," *Issues in Science and Technology* Winter 1991–1992:64.
22. Richard A. Houghton and George M. Woodwell, "Global Climate Change," *Scientific American* April 1989:10.
21. Andrew C. Revkin, "Cooling Off the Greenhouse," *Discover* January 1989:31.
23. Richard A. Kerr, "Iron Fertilization: A Tonic, but No Cure for the Greenhouse," *Science* 263 (1994):1089–1990.
24. Rudy Baum, "Researchers, Other Experts Examine Climate Engineering Issues," *Chemical & Engineering News* 7 March 1994:26–27.
25. COSEPUP, 76–78.
26. Gore, 240.
27. William A. Nierenberg, "Science, Policy, and International Affairs: How Wrong the Great Can Be," *Environmental Conservation* 20 (1993):197.
28. William D. Nordhaus, "Economic Approaches to Greenhouse Warming," in *Global Warming—Economic Policy Responses*, ed. Rudiger Dornbusch and James M. Poterba (Cambridge, MA: MIT Press, 1991), p. 42.
29. William D. Nordhaus, "An Optimal Transition Path for Controlling Greenhouse Gases," *Science* 258 (1992):1316.
30. Nordhaus, "Economic Approaches," 39–46.
31. Warren T. Brookes, "The Global Warming Panic," *Forbes* 25 December 1989:98.
32. Newton, 89.
33. Bierbaum and Friedman, 58–59.
34. Terrell J. Minger, ed., *Greenhouse Glasnost* (New York: The Ecco Press, 1990), p. 42.
35. Clinton, 1.
36. S. Fred Singer, ed., *Global Climate Change: Human and Natural Influences* (New York: Paragon House, 1989).
37. Brookes, 98.
38. Maurice Strong, "U.S. Gets a Stern Warning on Eve of Rio Conference," *San Francisco Chronicle* 3 June 1992:A2.

39. "Environmental Protection or Imperialism?" *Nature* 363 (1993):657.
40. "Policy Implications of Greenhouse Warming," Hearing before the Committee on Commerce, Science, and Transportation, United States Senate (Washington, DC: U.S. Government Printing Office, 1991), pp. 46–48.
41. Bolin, 94.
42. Brookes, 102.
43. Philip H. Abelson, "Chlorine and Organochlorine Compounds," *Science* 265 (1994):1155.

PART II

Computer Climate Models and the Earth's Temperature

FOUR

Climate Models

There was once a Zulu woman who had a deformed and very ugly baby. Because the baby was not acceptable to her tribe, she took him to a deep cave and raised him there.

One day as the boy gazed at some iron ore in the cave, it melted and turned into iron. The boy made a flying robot with powerful teeth and jaws. He built many more robots and set out to conquer the world with his army.

In his new world, the robots did everything for the people. Before long the people grew accustomed to this new way of life. Their lives were easy and comfortable, and they forgot how to do anything for themselves, even how to have children.

But the Tree of Life and the Earth Goddess were outraged, and they wiped out the evil youth and his robots, leaving no trace of his regime.

—*Zulu myth*[1]

The moral of the Zulu myth is that people should not place too much confidence in computers or computer models, thus abdicating their own responsibilities and their ability to make decisions. Computers have enormous benefits as tools of society, but people sometimes forget the limitations of computers and run the risk of relying on them even when they are wrong. The dire warnings about global warming and its consequences and the resulting policy recommendations and controversies are primarily based on the predictions of computer climate models.

What exactly is a computer climate model? How precise are the computer climate models at predicting the earth's weather for the near future? How accurate are they at predicting global climate at some point five, ten, twenty, or one hundred years from now?

COMPUTER MODELS

A computer model is a mathematical equation or series of equations that can be solved to predict the results of a real-life activity or process. These equations must include every factor that has an effect on the activity, and the input data must be representative of each factor. If the equations are correct, the predictions should be reliable.

Let us construct a simple computer model of a marathon runner as an example. Let's say we have measured the runner's average speed to be 5.0 miles per hour during the course of the 20-mile marathon—this is the input data. Using the equation

$$\text{Distance} = \text{Time} \times \text{Speed}$$

we can calculate that he or she will run 20.0 miles in 4.0 hours:

$$\text{Distance} = 20 \text{ miles} = 4.0 \text{ hours} \times 5.0 \text{ miles/hour}$$

We can "program" this equation into the computer and create the simplest of computer models for a marathon. With this information we can now predict how long the marathon will take, and if we know the speed of each runner, we can even predict who will win the race. We can also predict a multitude of other things.

What if the next race is a 100-mile supermarathon? Based on our model, the marathon runner would have run for

$$100 \text{ miles} \div 5.0 \text{ miles/hour} = 20.0 \text{ hours}$$

Wait! Is it realistic to think that a person can run a 100-mile race at the same average speed as a 20-mile race? The supermarathon is five times as far. For a race that lasts for 20 hours or more, our runner has to rest and eat. Let's face it—no one can run 100 miles at the same average speed that he or she can run 20 miles. The input data may be good for the duration of the short marathon, but can it be extrapolated to a 100-mile race? The answer is no! We must use realistic constraints.

It may be possible to extrapolate outside of our input data (the average speed during the 20-mile marathon), but this must be done with caution. It would be realistic to use the computer model for a 25-mile race, but probably not for a 50-mile one.

Consider all the factors that might influence the speed of our marathon runner:

- Conditioning
- Health
- Mental attitude
- Terrain
- Altitude
- Weather conditions (temperature, wind speed and direction, humidity, etc.)

In order for us to have an accurate marathon computer model, we must include a precise, accurate mathematical equation for each of these factors. Is this possible? No! Can we come close? Maybe.

The point of this example is to stress that a model is only as good as the equations are accurate at describing the activity, and then it is only as good as the constraints are realistic. For example, the modeler should not permit the length of the race to be unrealistic. The further outside the time frame of data collection that we try to extrapolate and predict results, the greater the uncertainties associated with those results will be.

GLOBAL CLIMATE MODELS

A global climate model is much, much more complex than that described for our marathon runner, and therefore it is far more difficult to calculate with certainty! A typical modeling approach is to divide the surface of the earth into equal areas, identify several regions of the atmosphere above each of these areas, and calculate the climate parameters for each of the resulting volumes, or "boxes," of atmosphere as a function of time, say every thirty minutes or so. Stephen Schneider, at the National Center for Atmospheric Research, who has been prominent in developing computer climate models, has described this approach using a global grid of 4.5° latitude by 7.5° longitude. This grid represents an area 500 by 640 kilometers (roughly 310 by 400 miles), which is about the size of New Mexico. The atmosphere above the grid is divided into nine segments, creating 17,280 volumes of atmosphere that must be determined for all of the climatic parameters for each time period. If the calculations are performed for a thirty-minute time interval, Schneider figures that it requires about 100 hours of computer calculations for a ten-year climate simulation using a Cray XMP supercomputer. Computer time costs less with each passing year, but in 1989, the cost was approximately $1,000 per hour, according to Schneider.[2] This means that a 100-year computer simulation would cost $1 million! Some computer experiments run simulations for up to 500-year climate forecasts. The costs of modeling climate are not insignificant!

William A. Nierenberg, Director Emeritus of Scripps Institution of Oceanography at La Jolla, California, states that the United States is currently spending over $1 billion per year on global warming research and computer modeling![3] These research costs must be considered as part of the overall costs of global warming.

To understand the many aspects of global climate models it is necessary to know the following basic definitions:

- *Global climate model*: A general term describing any model that attempts to simulate the earth's climate. (Most models are given a more specific name such as those described below, and many are given additional acronyms that associate them with a specific organization.)

- *Convective model:* A model that seeks to balance the energy input from the sun with all of the energy outflow from the earth. An imbalance in the energy will result in either warming or cooling.
- *General circulation model (GCM)*: This is a model that attempts to describe the circulation of the atmosphere (or the ocean) in terms of all the climatic parameters, such as wind speed and direction. These models are three-dimensional and include altitude data as well as latitude and longitude. There are at least nineteen different atmospheric GCMs and several different oceanic GCMs in current use by scientists. The major difference between the various models is the approach that each uses for handling the dynamics of water vapor, cloud cover, solar energy, and the interface to the oceans.
- *Coupled model:* When an atmospheric GCM is combined with an oceanic GCM to simulate more realistically the atmospheric–oceanic interface, the result is called a "coupled model."

In all cases the models use actual weather data as input to "calibrate" the system. This input data usually includes all the required weather parameters over a twenty- to fifty-year time frame. In some cases, features (or parameters) of the weather, such as cloud formation or cloud cover, are calculated based on other parameters, such as relative humidity. Such equations are called "parameterizations" of the clouds.

To see just how complicated models become, we must understand the difference between weather and climate, as these terms are defined and used by climatologists, the scientists who study the weather. Schneider has defined climate as follows:

> Climate is typically the average state of the atmosphere observed as the weather over a finite time period (e.g., a season) for a number of different

> years. Thus, we can speak of the climate of a day-night cycle, month, season, year, decade, or even longer period. Climate is usually defined by the mean state together with measures of variability or fluctuations.[4]

In other words, the climate is the average of the weather over a particular time period. In order to predict the global climate, we must be able to predict the weather!

VARIABLES AND UNKNOWNS

What information must be included in the computer models to ensure that a realistic description of the weather can be accomplished? Lennart O. Bengtsson of the Max Planck Institute for Meteorology in Hamburg, Germany, has summarized the variable parameters that must be considered in a complete climate model:

- Variables important for the description of the general circulation, including atmospheric pressure, wind speed and direction, and temperature throughout the depth of the atmosphere. Daily, weekly, and seasonal variability must be addressed. This is the day-to-day weather.
- Variables critical for energy transfer, equilibrium, and the hydrological cycle. These include the interface between the oceans and the atmosphere, the oceans and the ice packs, and the oceans and the landmasses, and must take into account the complex variation of the water vapor, cloud, and precipitation interactions.
- Variables important for climate feedbacks, such as the radiative effects of clouds, snow cover, soil moisture, aerosol concentrations, greenhouse gas concentrations, and sea ice.
- Variables of extreme events—cyclones, hurricanes, and volcanoes.[5]

Detailed records for many of the everyday weather variables are now kept at weather stations all over the world. But most of the other variables (the energy variables, the climate-feedback variables, and the extreme-event variables) have been ignored or oversimplified by computer modelers. In some cases this is because the variables were perceived to be essentially constant, such as the solar constant; in other cases they were thought to be unimportant in determining the long-term climate, such as the mix of farmland versus forests on the landmasses; and finally, in some cases, the scientists simply do not know enough about the variable to include it properly into the model. This is especially true in the case of the hydrological cycle, as discussed in Chapter 8.

For long-term results, there are many known weather cycles that should be evaluated. For example, to determine the climate for the month of January, climatologists would observe the weather each January for several (preferably many) years and calculate the averages. These averages would then represent the climate for January. The obvious cycle that affects the average temperature is the day–night (diurnal) cycle—the climate will be moderated because it is usually colder at night than during the day, and measurements for both the day and night temperatures would be included in calculating the average. If only the daytime climate is of interest, then the observations must be altered to include daytime temperatures only. For the greatest accuracy, the climatologist must include the fact that in the Northern Hemisphere the days are getting longer throughout the month of January.

Following this line of reasoning, in order to look at the general climate for a longer period, say a decade, the climatologist must realize that there are other longer-term cycles to deal with, such as solar cycles, orbital variations, and El Niño, a major ocean cycle that affects global weather. This brings up a very important point. A climatologist cannot use observations of the current or past weather to predict the future unless the computer model considers all cyclical variables and noncyclical variables accurately.

An excellent example of the problems associated with extrapolation of current data to future predictions lies with the effects of great natural events such as volcanoes. It is well known that a major eruption from a volcano can cause observable weather changes for months, sometimes years (see Chapter 9). Unfortunately, geologists cannot yet predict when a volcano will erupt, nor can they predict how large the eruption will be. The best climatologists can do is to predict what the climate will be if there are no volcanic eruptions or other major natural events.

As if all the unknown problems of prediction weren't enough, a knowledge of the weather alone will not provide a satisfactory model for long-term climate prediction. The weather in our gaseous atmosphere cannot be isolated from the important interconnections that the atmosphere has with the earth itself. The oceans play a powerful role with respect to the extent and nature of the winds and moisture in the atmosphere, both of which affect the temperature and the type and extent of clouds in the atmosphere. A specific example is that of the El Niño event, which is associated with the change in wind direction and ocean temperature off the coast of Peru. It is thought that this isolated and localized El Niño event causes both droughts and monsoons on several continents around the world because of its connections with the global climate.

All the weather features we have discussed have major influences on the climate, as do the nature and composition of the atmosphere (and the ocean for that matter). The composition of the earth's atmosphere is generally constant if the variation of water vapor is not considered.[6] However, the concentration of water in the atmosphere is not constant in any sense of the word. It varies from day to day, from day to night, from hour to hour, and from one part of the globe to another. Think about this: Water vapor is the major greenhouse gas! This means that the greenhouse effect will occur to a much greater extent over the moisture-rich part of the globe and likewise to a much lesser extent over the dry desert areas. Carbon dioxide is only 0.035% (about 350 parts per million) of the dry atmosphere, and all other "greenhouse gases" are even less concentrated, as we see in Chapter 6. Global warming due to the greenhouse gases is not a simple factor that can be applied in a simple way to the earth's climate.

THE EVOLUTION OF GLOBAL CLIMATE MODELS

The first scientific developments leading to a climate model were made by Vilhelm Bjerknes, a Norwegian physicist and meteorologist. He developed circulation theories applicable to the large-scale motions of the atmosphere (and ocean), which were published in 1904 in a paper titled "The Problem of Weather Prediction Considered from the Point of View of Mechanics and

Physics." The calculations required by his theory were too complex for practical everyday use in weather prediction at the time. Very slow progress was made until the advent of computers because of the requirement for immense numerical calculations in even the simplest models.[7] Of course, the results of the earliest models took much too long to calculate for the results to be applied in real time.

This field of study was made practical with the development of the computer and the great advances in weather measurement instruments during the second half of the twentieth century. Prior to the development of electronic measurement systems, much of the weather data was crudely acquired. There are now stations around the world—for example, at every national or international airport in virtually every country—using instrumentation with standardized calibration methods. These stations provide data that can be trusted to have the same meaning from station to station. Some 160 member countries and territories comply with the rules set forth by the World Meteorological Organization.

In the 1950s and 1960s scientists started to develop the general circulation models (GCMs) that are the backbone of today's climate models. The atmospheric GCMs made general simplifications with respect to any variation of the sun's energy or to effects caused by the ever-changing cloud cover and ocean circulation. These simplifications create major uncertainties associated with predicting the results of global warming in the years to come. There are currently at least nineteen GCMs that have been developed at major research centers in the United States, Europe, Russia, and Japan. As mentioned in Chapter 1, the results derived from the different GCMs do not agree with each other because of the variety of assumptions and simplifications used in the various models.[8] A specific example of this disagreement is discussed in the following section.

Much of the current research in climate modeling focuses on improvements in incorporation of these variables, both in obtaining the data and in modeling their effects. Of major interest are the coupled models that incorporate both the oceanic circulation and atmospheric circulation models to obtain a more realistic simulation of the atmosphere–ocean interface. Due to the increased complexity of the computer models, the data from these enhanced models often provide unrealistic predictions and take many iterations of refinement before the results become meaningful. Improving the GCMs is a process of taking many little steps.

The most recent climate assessment of the International Panel on Climate Change (IPCC) included results of coupled atmospheric and oceanic GCMs, but these simulations held solar variation and cloud dynamics as essentially constant. But as stated in Chapter 1, the predicted global warming results varied by over 300%, based on a doubling of the carbon dioxide levels in the atmosphere.

UNCERTAINTIES IN COMPUTER CLIMATE MODELS

Much discussion has taken place regarding the level of uncertainty associated with the predictions of computer climate models. However, the popular position taken by many of the environmental groups and some scientists seems to be that drastic action should be required in spite of these uncertainties. Let's take a further look at the level of uncertainty in the computer models.

Dr. Stanley Grotch of the Atmospheric and Geophysical Sciences Division at the Lawrence Livermore National Laboratory performed a statistical analysis of four atmospheric GCMs using an empirical (observed) data base collected from 1957 to 1973. Using the same calibration data for all models, he compared the results to see if these different models could predict the same details of climate. To do this, he plotted the values predicted by one model on one axis of a graph versus the values predicted by another model on the other axis.[9] If the models predict climate details in the same way, the plot will be a straight line with a 45° angle (see Figure 4.1). Figure 4.2 shows Grotch's plots of small-area temperature predictions for all possible pairings of the four models in Grotch's comparisons.

As you can see, there is nothing resembling a straight line with a 45° angle. In fact, we could draw a line at almost any angle. This tells us that there is absolutely no correlation between the predictions of the four models!

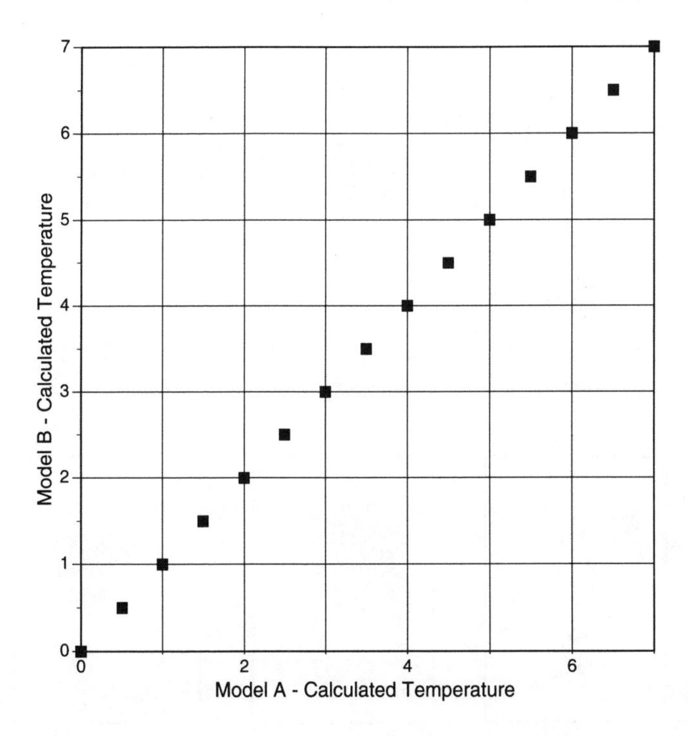

FIGURE 4.1 Ideal correlation for a crossplot.

Remember that the four different models were calibrated using the exact same set of weather data, and yet they could not predict the same weather results on a small regional scale!

No one can refute the importance of predicting the weather. If the people in the midwestern United States had only known that it would rain for "forty days and forty nights" and more during the summer of 1993, they might have been able to mitigate the billions of dollars of damage to both property and crops as well as the effect of these losses on thousands of people. The development of better climate models for short- and medium-term weather prediction is vitally important.

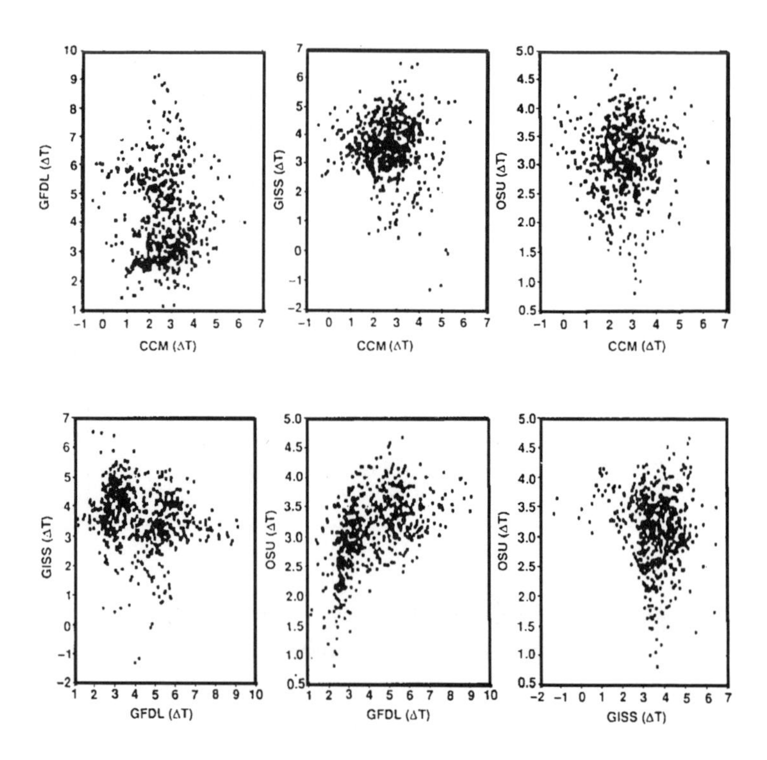

FIGURE 4.2. Crossplots of predicted ΔT (change in temperature) for four GCM models. These plots compare the predicted ΔT for all model pairs plotted one against the other at the same 4° × 5° gridpoints. Land gridpoints over the range 62° S to 62° N are included. The axis scales vary between plots to show the scatter of the data more clearly. These figures are taken from Stanley L. Grotch, "A Statistical Intercomparison of Temperature and Precipitation Predicted by Four General Circulation Models with Historical Data," in *Greenhouse-Gas-Induced Climatic Change: A Critical Appraisal of Simulations and Observations*, ed. M. E. Schlesinger (New York: Elsevier, 1991), p. 12. The models are identified as follows: CCM, Washington and Meehl, *J. Geophys. Res.* 88 (1984):6600–6610; GFDL, Manabe and Wetherald, *J. Atmos. Sci.* 44 (1987):1211–1235; GISS, Hansen et al., *Climate Processes and Climate Sensitivity* (Washington, DC: American Geophysical Union, 1984), pp. 130–163; OSU, Schlesinger and Zhao, *J. Climate* 2 (1989):459–495.

MODEL VALIDATION OR CONFIRMATION

When a model is developed, experiments must be performed to give the modeler confidence that the results of the model provide realistic information. This is done in a variety of ways. A common approach is to calibrate the model with an empirical data set that covers only part of the information that is known and use the model to predict information about the rest of the data set, which was not used in the calibration. For example, if the known weather data covers the years from 1900 to 1990, the modeler might use the weather data from 1940 to 1990 as calibration data, then use the model to predict the weather during the period from 1900 to 1940. If these predictions match the observed weather from this period, the modeler can feel confident that his or her model is realistic. This is a simplified statement of validation and there are many variations and subtleties in validation procedures.

You probably have not read about model validation in any of the popular literature. The omission of discussion about validation stems from the fact that current computer climate models cannot be validated by these techniques. That is, they do not accurately predict known weather on a global basis.

Dartmouth scientists Naomi Oreskes and Kenneth Belitz, and Kristin Shrader-Frechette, a philosopher from the University of South Florida, have discussed this issue in a *Science* article titled "Verification, Validation, and Confirmation of Numerical Models in the Earth Sciences," in which they state: "Verification and validation of numerical models of natural systems is impossible." They explain that a natural system is an open-ended system in which all of the inputs can never be fully expressed by an equation. They argue that only closed systems can be fully modeled. A closed system may be defined, very simplistically, as an equation that can be solved absolutely and irrefutably with the input data. They point out specifically that the hydrological cycle is an example of an open-ended system. In their opinion a model of a natural system can only be confirmed by the procedure described in the preceding paragraphs.[10] This type of validation requires that a portion of the input data be used for calibration of the model and that the model then be used to predict known events accurately within the time frame used for calibration.

WHAT DO THE CLIMATE SCIENTISTS SAY ABOUT THE MODELS?

Many climatologists in the computer modeling field have a healthy appreciation for the weaknesses and shortcomings of their models and, in their technical publications, are careful to point these out. This applies to inadequate input data as well as to the uncertainty of prediction from the models. A sample of these cautionary statements is provided below.

Roy L. Jenne, of the National Center for Atmospheric Research (NCAR) at Boulder, Colorado, stated the consensus of many climatologists when he wrote in 1991:

> If we assume that the global surface warming over the previous century is about 0.4°C, the next question is whether it is caused by the greenhouse effect. Many historical and paleoclimatic studies have been made that give a considerable amount of evidence for the climate over the last 1000 years and much longer. These studies are based on data such as tree-ring growth, harvests of grapes and other crops, pollen deposits, and various historical documents. It appears that during the past five or ten centuries the climate showed periods of warming and cooling that are not unlike what we have measured in the past century. This climate variability happened without a greenhouse effect caused by the activities of mankind. Even long runs of climate models (with no change in greenhouse gases) show a natural variability. Therefore, on the basis of observations, the present warming cannot be said to be caused by the greenhouse effect.[11]

In 1984 when the interest in global climate changes due to greenhouse gases was rapidly increasing, John T. Houghton (Meteorological Office, Bracknell, UK) edited a major work that summarized the state of the art of climate modeling and described The World Climate Research Programme, an international program, focused on the study of the earth's climate. In it he and Pierre Morel of the World Meteorological Organization, Geneva, Switzerland, wrote:

> Estimates with current models of possible climatic changes from various causes, both so far as global average effects and the regional variations are concerned, need to be treated with great caution, however, because the models do not yet include many of the important feedback processes.[12]

Remember that the models described in this reference were the ones on which the IPCC made its conclusions and recommendations, and Houghton is one of the editors for IPCC reports.

C. E. Leith, also of NCAR, emphasized the modeling problems, especially with respect to cloud and water vapor issues, when he summarized the main sources of concern with the research on global climate:

> The greatest source of uncertainty in GCM climate sensitivity experiments at present arises from the crudeness of the cloud–radiation interaction and feedback processes in the models.
>
> This vertical advection problem [crudeness with dealing with clouds] also renders dubious any model calculation of water vapor in the stratosphere. Unfortunately, radiative balance in the stratosphere is dependent on its water vapor content which is often specified rather than computed on the assumption that the measurements, crude as they are, are more reliable than a model calculates.[13]

The important role that the oceans play in the earth's climate had been greatly oversimplified in the atmospheric GCMs, and Leith had serious concerns about this simplification. He goes on to state: "The ocean models must be coupled with the atmospheric models for GCM to have any accuracy."[14]

In 1991 an important publication appeared: "Greenhouse-Gas-Induced Climatic Change: A Critical Appraisal of Simulations and Observations," edited by Michael E. Schlesinger, Department of Atmospheric Sciences, University of Illinois. This work describes research during the mid-1980s and early 1990s and provides a scientific critique of the models as they stood at that time. Following Leith's admonishment to couple the atmospheric GCMs to the ocean models, several research programs have moved in that direction. One example is the work of Dr. Ulrich Cubasch described below.

Unfortunately, it is much easier to discuss the combination of extremely complex computer models than it is to accomplish the merger. In Schlesinger's book, Ulrich Cubasch, from the Max Planck Institute for Meteorology, discusses the research at that institute in combining an atmospheric GCM with an oceanic GCM:

> A low-resolution version of the . . . global atmosphere model has been coupled to a global ocean model developed at the Max Planck Institute in Hamburg. . . . Even though each model reaches stationarity [steady state] when integrated on its own, the coupling of both creates problems. . . . In the coupled experiment the combined ocean–atmosphere system drifts toward a colder state. To counteract this problem, a flux correction has been applied which balances the mean biases of each model. This method almost eliminates the climate drift of the coupled model.[15]

What is meant by a "flux correction?" This is a mathematical factor determined by the modeling scientist to make the predictions, or model calculations, behave the way he or she believes they should. Generally, such corrections do not have fundamental meaning, and they can be arbitrary (the term "fitting the data" was used in Chapter 2). Without the "flux correction," this model predicted global cooling!

In another important reference on climate modeling, Lennart Bengtsson, also from the Max Planck Institute, further discusses the coupling problem:

> However, when the systems are coupled, one normally discovers that the coupled systems slowly drift into another state which is far from the observed climate. The drift clearly implies that the original subsystems contain errors which were too small to be detected, or which were compensated for in some way.[16]

Michael C. MacCracken from Lawrence Livermore National Laboratory and John Kutzbach from the Center for Climatic Research at the University of Wisconsin concur:

> The rising concentrations of carbon dioxide and other greenhouse gases are projected to warm the global climate by a few degrees Celsius over the next century. The numerical models upon which these estimates are based, however, have many limitations, including especially their approximations in representing physical processes and their coarse spatial resolution. Therefore, results from these models cannot yet be used to make definitive assessments of potential ecological and societal impacts

> or to develop adaptation strategies in important areas such as agriculture and water resources.[17]

It is important to understand that atmospheric scientists have, with few exceptions, maintained their open-mindedness and scientific objectivity with respect to what can and cannot be concluded from computer climate models. The results of the models have been misused and misstated by environmental advocates to scare the public into action. There may be good reason for action, but action should *never* be taken because of misstated scientific results.

Caution was further emphasized by Jim Hansen and a research team from NASA's Goddard Institute for Space Studies in 1991:

> The conventional wisdom is that it is not yet possible to obtain reliable conclusions about regional climate impacts, principally for two reasons. First, the representations of atmospheric and surface processes in current climate models are highly simplified, and the models show a very wide range in their predictions for climate change at any particular region. Second, none of the existing climate models simulates the ocean realistically, and changes in ocean currents could alter regional climate.[18]

This group has nevertheless advocated drastic and immediate action by our government in spite of the inconclusive nature of the models.

It is clear from the statements and conclusions of climatologists currently active in the computer climate modeling field that they cannot accurately predict climate for even a few years, much less decades into the future.

PREDICTING THE GREENHOUSE WARMING EFFECT

Most of the global warming model predictions project a temperature rise of 1.5–4.5°C (2.7–8.1°F) due to increasing greenhouse gases. These predictions are based on the following procedure: First, climatologists calibrate their model with an empirical data set measured for a specified duration. They run, or integrate, the model to simulate global temperature during a prescribed time, say 50 or 100 years, to provide a "baseline" temperature variation. Then they double the carbon dioxide concentration (the input data) and rerun the model. Finally, they subtract the temperature results of the first run from those of the second. A temperature change is obtained in this manner for each of the areas that the modeler set up for his grid. A 4.5° latitude by 7.5° longitude grid was discussed earlier in this chapter as an example of a typical GCM grid (about 310 by 400 miles in size). The average, or mean, temperature change from all the temperatures calculated for the whole globe (about 2,000 different temperatures) is thus reported as the global temperature change due to the greenhouse effect.[19] These calculations have mostly been made with atmospheric GCM models that do not include ocean dynamics, cloud dynamics, solar dynamics, aerosols, or adequate water vapor dynamics.

This is not a realistic way to treat an increase in carbon dioxide or any other greenhouse gas. Humanity has not and will not suddenly dump twice the amount of greenhouse gases into the atmosphere. These gases are slowly increasing; they have increased about 25% in the past century. The worst-case scenarios would have the greenhouse gases doubling in about 50 years, but more moderate scenarios indicate that it would require 75–100 years for this doubling to happen.

A realistic model would slowly increase the concentrations of these gases over the years as they would increase in the real world. The practice of arbitrarily doubling the concentration of one atmospheric component, such as carbon dioxide, will cause an unrealistic disruption in the process. Climatic interactions are too complex for a jump in a particular variable to have true meaning. According to the IPCC, several recent experiments that have followed a more realistic approach of gradual gas buildup in the atmosphere predict global temperature increases of only about one-third of those models that double the CO_2 concentration.[20] This doubling is only one of the many flaws in the current methods of predicting global warming, but perhaps one of the most important.

CHAOS

A final point should be made concerning the chaotic behavior of global climate. The scientific concept that is referred to as chaos was identified by Ed Lorenz, a meteorologist at MIT, who was studying numerical models associated with atmospheric circulation. Lorenz published the seminal work on chaotic behavior in the *Journal of Atmospheric Sciences* in 1963. In the intervening decades, much has been learned about the chaotic behavior of systems, such as the atmospheric and oceanic circulation systems, which are nonlinear as are many phenomena in nature. A linear system is one in which behavior can be described by the simple mathematics of a linear equation, in which relationships can be expressed by a straight line on a graph. Linear equations are solvable; if you know the input data, you can calculate a specific solution. Nonlinear systems cannot be so simply treated; they are not solvable. One of the most important traits of a nonlinear system is its sensitivity to initial conditions. A small difference in the initial input conditions can make a tremendous difference in the prediction of conditions at some later time. This means that small differences in the input to a GCM could mean major differences in the final prediction as to the amount of global warming. The longer the time frame of the prediction, the larger the error in temperature prediction will be. According to Peter Read, it would take 2,500 years of daily temperature data measured to a precision of ±0.01°C (±0.0056°F) to establish whether or not the natural weather is indeed chaotic;[21] however, it is well known that many aspects of the climate system are nonlinear in their behavior. This means that there is probably a chaotic component to climate systems.

TOWARD A MORE REALISTIC APPROACH

Climatologists are turning in the direction of a more realistic approach. They are including more realistic ocean, cloud, and aerosol dynamics in their computer models. Unfortunately, the costs of computer time go up tremendously when all these factors are included. Although current studies must be considered preliminary, the indications are that when ocean dynamics are included in climate models, the increase in predicted global warming is moderated and not as dire as previous predictions indicate.[22] In fact, the predicted global warming has tended downward as greater improvements have been made to climate models.

The enormity of creating a computer model that simulates the earth's climate cannot be overemphasized. The scientists who have devoted their careers to this research are dedicated and sincere, and they are making considerable progress. But the limitations and shortcomings of current models must be kept in perspective. A summary of these is compelling:

- Computer global warming models are extremely complex, costly, and time consuming.
- These are the same computer models that are used to predict the weather—which cannot be accurately predicted for up to four or five days.
- There are variables and unknowns that have not been included adequately in current computer models:
  - Long-term cyclic processes and natural major events, such as volcanic eruptions or the El Niño
  - Ocean, solar, cloud, and water vapor dynamics
- When complexities such as ocean dynamics are figured into the models, they often predict unrealistic drift, which can be in the wrong direction for global warming. They can predict global cooling! (So-called flux corrections have been used to make the drift more "realistic" for the climatologist. These corrections are not based on fundamental theory and can be arbitrary.)
- Atmospheric GCMs (there are at least nineteen different ones) do not predict small-scale events with any accuracy, nor do the predictions correlate with one another even using the same input data.
- The increases of carbon dioxide have not been realistically handled in atmospheric GCMs except in very few experiments, and when the increase has been applied slowly, a much lower rate of global warming is predicted.
- Efforts to validate or confirm computer models for periods with known empirical data have failed.

- The observed natural variation of the earth's temperature is as great or greater than the predicted variation of computer models.
- The scientists working in the field today are in general agreement that computer climate models are not adequate to provide proof of global warming.

Dr. Reid Bryson, Senior Scientist at the Center for Climatic Research, University of Wisconsin, sums up our knowledge about computer models as follows:

> The fact that the large computer models indicate such a temperature rise as a consequence of increased carbon dioxide cannot be taken as evidence of truth; for any such model is merely a formal statement of the modeler's opinion of how the atmospheric system works.[23]

Dr. Bryson is very perceptive; a preconceived belief in the outcome of a prediction will lead some scientists to adjust the computer to achieve the "right" results. The models are so complex that it is very easy to rationalize the addition of flux correction, for example. This comes under the same heading as the self-fulfilling prophecy. One must treat the results on models at this level of complexity with a great deal of skepticism until they have been thoroughly validated.

REFERENCES

1. Malcolm E. Weiss, *Gods, Stars, and Computers* (Garden City: Doubleday, 1980), pp. 113, 117.
2. Stephen H. Schneider, *Global Warming: Are We Entering the Greenhouse Century?* (San Francisco: Sierra Club Books, 1989), p. 297.
3. William A. Nierenberg, "Science, Policy, and International Affairs: How Wrong the Great Can Be," *Environmental Conservation* 20 (1993):196.
4. S. H. Schneider, "Introduction to Climate Modeling," in *Climate System Modeling*, ed. K. E. Trenberth (Cambridge: Cambridge University Press, 1992), p. 6.
5. L. O. Bengtsson, "Climate System Modeling Prospects," in *Climate System Modeling*, ed. K. E. Trenberth (Cambridge: Cambridge University Press, 1992), p. 707.
6. K. E. Trenberth, ed., *Climate System Modeling* (Cambridge: Cambridge University Press, 1992), p. xxix.

7. J. J. Hack, "Climate System Simulation: Basic Numerical & Computational Concepts," in *Climate System Modeling*, ed. K. E. Trenberth (Cambridge: Cambridge University Press, 1992), p. 284.
8. Committee on Science, Engineering, and Public Policy (COSEPUP), *Policy Implications of Greenhouse Warming* (Washington DC: National Academy Press, 1991), p. 108.
9. S. L. Grotch, "A Statistical Intercomparison of Temperature and Precipitation Predicted by Four General Circulation Models with Historical Data," in *Greenhouse-Gas-Induced Climatic Change: A Critical Appraisal of Simulations and Observations*, ed. M. E. Schlesinger (New York: Elsevier, 1991), pp. 3–16.
10. Naomi Oreskes, Kristin Shrader-Frechette, and Kenneth Belitz, "Verification, Validation, and Confirmation of Numerical Models in the Earth Sciences," *Science* 263 (1994):641–646.
11. Roy L. Jenne, "Climate Trends, the U.S. Drought of 1988, and Access to Data," in *Greenhouse-Gas-Induced Climatic Change: A Critical Appraisal of Simulations and Observations*, ed. M. E. Schlesinger (New York: Elsevier, 1991), p. 198.
12. J. T. Houghton and P. Morel, "The World Climate Research Programme," in *The Global Climate*, ed. J. T. Houghton (Cambridge: Cambridge University Press, 1984), p. 2.
13. C. E. Leith, "Global Climate Research," in *The Global Climate*, ed. J. T. Houghton (Cambridge: Cambridge University Press, 1984), p. 18.
14. Ibid., 18.
15. U. Cubasch, "Preliminary Assessment of the Performance of a Global Coupled Atmosphere-Ocean Model," in *Greenhouse-Gas-Induced Climatic Change: A Critical Appraisal of Simulations and Observations*, ed. M. E. Schlesinger (Amsterdam: Elsevier, 1991), p. 137.
16. Bengtsson, 719.
17. M. C. MacCracken and J. Kutzbach, "Comparing and Contrasting Holocene and Eemian Warm Periods with Greenhouse-Gas-Induced Warming," in *Greenhouse-Gas-Induced Climatic Change: A Critical Appraisal of Simulations and Observations*, ed. M. E. Schlesinger (New York: Elsevier, 1991), pp. 17–18.
18. J. Hansen, "Regional Greenhouse Climate Effects," in *Greenhouse-Gas-Induced Climatic Change: A Critical Appraisal of Simulations and Observations*, ed. M. E. Schlesinger (New York: Elsevier, 1991), p. 212.
19. COSEPUP, 111–116.
20. J. T. Houghton, G. J. Jenkins, and J. J. Ephraums, eds., *Climate Change: The IPCC Scientific Assessment* (Cambridge: Cambridge University Press, 1990), pp. 15–16.
21. Peter L. Read, "Applications of Chaos to Meteorology and Climate," in *The Nature of Chaos*, ed. Tom Mullin (Oxford: Oxford University Press, 1993), p. 226ff.
22. W. M. Washington and G. A. Meehl, "Characteristics of Coupled Atmosphere-Ocean CO_2 Sensitivity Experiments with Different Ocean Formulations," in

Greenhouse-Gas-Induced Climatic Change: A Critical Appraisal of Simulations and Observations, ed. M. E. Schlesinger (New York: Elsevier, 1991), p. 667.

23. Reid A. Bryson, "Simulating Past and Forecasting Future Climates," *Environmental Conservation* 20 (1993):339.

FIVE

The Earth's Temperature

I wish to tell you about a marvelous way in which I am accustomed to measure, with a certain glass instrument, the cold and hot temperature of the air of all regions and places, and of all parts of the body; and so exactly, that we can measure with the compass the degrees and ultimate limits of heat and cold at any time of day. It is in our house at Padua and we show it very freely to all. We promise that a book about medical instruments that are not well known will shortly appear, in which we shall give an illustration of this instrument and describe its construction and uses.

—COMMENTARIA IN ARTEM MEDICINALEM GALENI, PART III SANTORIO, 1612[1]

This passage by Santorio is possibly the first written mention of a glass thermometer. The development of the thermometer and the science of temperature measurement provided a means for measuring the temperature of the atmosphere.

The currently popular view of global warming holds that during the past century the earth's temperature has been increasing because of the increasing level of greenhouse gases in the atmosphere due to man's activities. The International Panel on Climate

Change (IPCC) has concluded that this temperature increase is between 0.3 and 0.6°C (0.5–1.1°F), but they add the caveat that this level of change is within the earth's natural variation. However, the figure of 0.5°C (0.9°F) has been generally accepted by government policymakers, environmentalists, and the media as the anthropogenic increase.

The thought of this very small change in temperature over such a large time span brings to mind many questions:

- How can one differentiate such a small temperature change over the immense time period of 100 years when the day-to-night temperature change is often 30–40°F, with an even greater seasonal variation?
- How is the average or mean temperature of the whole planet calculated?
- How was the temperature measured 100 years ago?
- What is the natural variation of the earth's temperature?

The task of obtaining the earth's temperature seems formidable. Let's look at what we know about the temperature of the earth during the past, at the historical and current methods used to measure the temperature, and see just how confident we can feel about their accuracy.

THE EARTH'S TEMPERATURE THROUGH HISTORY AND THE MILANKOVITCH THEORY

Scientists have long been intrigued with the variation of temperature on the earth. And, indeed, they are aware of ages that were much colder than today, when glaciers spread over most of Europe and North America. There are indirect records provided by tree ring and ice core studies that provide information about these past temperature cycles and the great ice ages. Figure 5.1 provides the big picture with respect to the earth's temperature history. The large temperature swings from ice age to interglacial took place over a very long time, and it was more often cold than

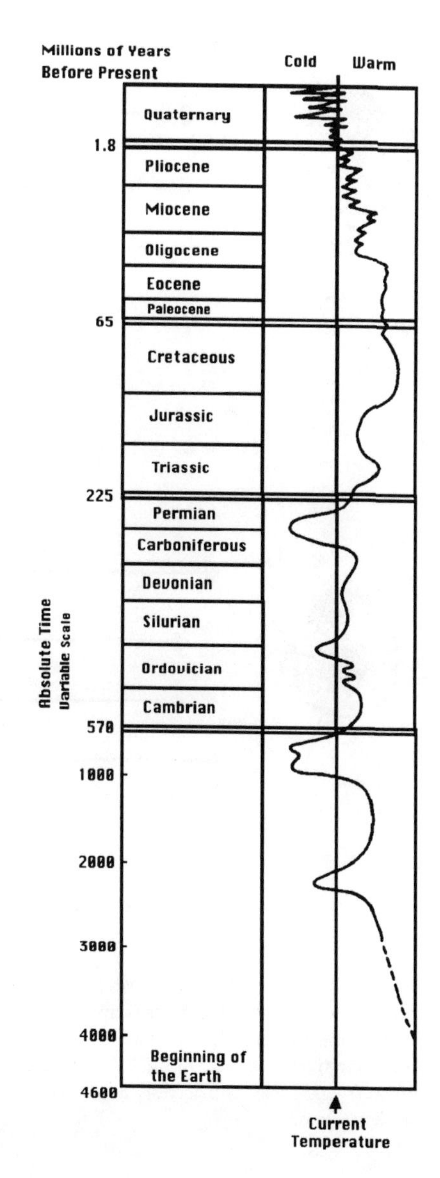

FIGURE 5.1. History of the earth's temperature. The figure depicts the variation in the earth's temperature from about four billion years ago to the present. Notice that for much of this time the earth was much warmer than during the past million years or so. This figure was adapted from data in Melvin A. Bernarde, *Global Warning . . . Global Warming* (New York: John Wiley & Sons, 1992), p. 26.

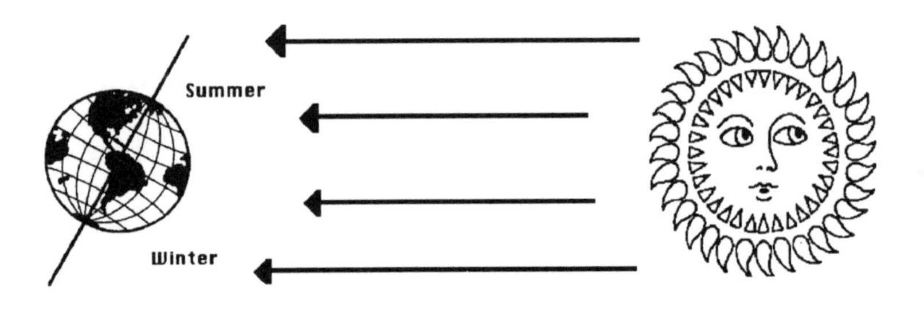

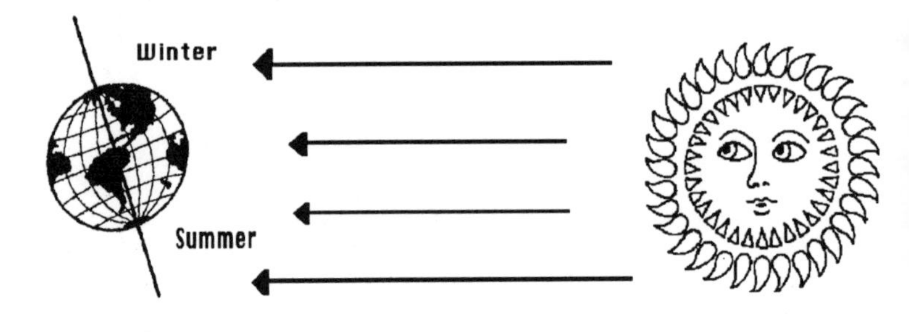

a

FIGURE 5.2. Variation in the earth's orbit about the sun: a, the tilt of the earth's axis, which is responsible for the seasons; b, the precession of the earth's rotation about its axis and the wobble in its precession; c, the gradual change of earth's orbit from more circular to more elliptical (eccentricity). These variations follow 20,000- to 100,000-year cycles and account for the ice age–interglacial changes for the past two million years.

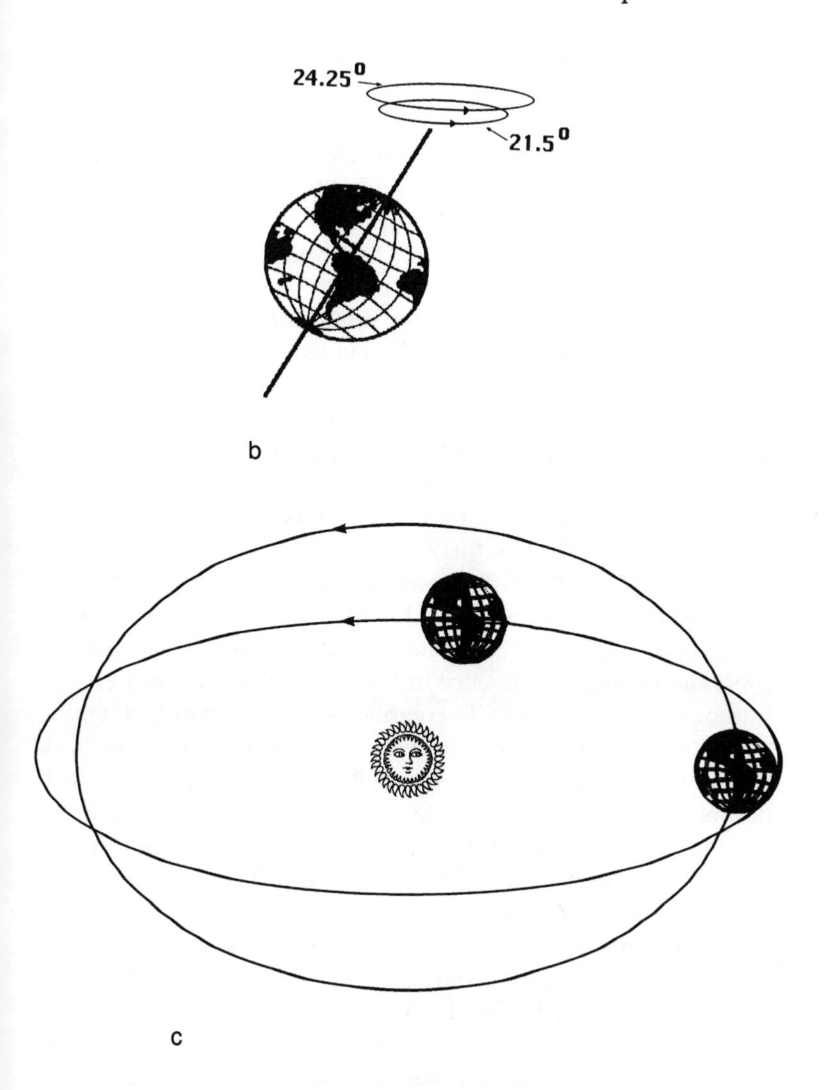

FIGURE 5.2. (continued)

warm. These long-term temperature variations are undoubtedly due to variations of the earth's orbit around the sun.

The orbital effects were first identified by a Serbian engineer, Milutin Milankovitch, who taught mechanics and theoretical physics at the University of Belgrade. His studies, published in 1930 and 1938, maintained that the ice ages were caused by the variation in solar energy reaching the earth that was a result of changes in the tilt of the earth's axis, precession of the earth's axis, and eccentricity of the earth's orbit about the sun over time. These phenomena are illustrated in Figure 5.2. The evidence for the Milankovitch theory is developed in Professor Melvin Benarde's excellent treatment on the "Reasons for Seasons."[2] As the earth rotates about its axis, the axis is tilted with respect to the sun, as illustrated in Figure 5.2a, and this is the cause for the seasonal changes from winter to summer. But the axis also rotates, which occurs over a very long time cycle of 20,000 years; this axis rotation is called precession and is illustrated in Figure 5.2b. There are wobbles or small variations in the axis precession, as the tilt slowly changes from 21.5° to 24.25° with a cycle of about 40,000 years. Both the wobble in tilt angle and precession cause a difference in the amount of the sun's energy impinging on the earth. To complicate the situation further, the earth's orbit changes from more elliptical to more circular and back again over the years. This eccentricity in the earth's orbit, which occurs on a 100,000-year cycle, causes the earth to vary in its distance from the sun, which in turn causes a variation in the amount of sunlight impinging on the earth. Thus the amount of solar radiation hitting the earth would not be a constant even if the sun were emitting a constant amount of radiation, which it is not. The sun's radiation varies with the sunspot cycle of about 11.1 years (the sunspot cycle is discussed in the next chapter). All of these effects must be considered if computer models are to be completely accurate.

During the past million years there have been ten ice ages, each lasting about 100,000 years, interspersed with short interglacial periods of relative warmth, lasting only about 10,000 to 12,000 years. All information prior to the past century or so is indirectly

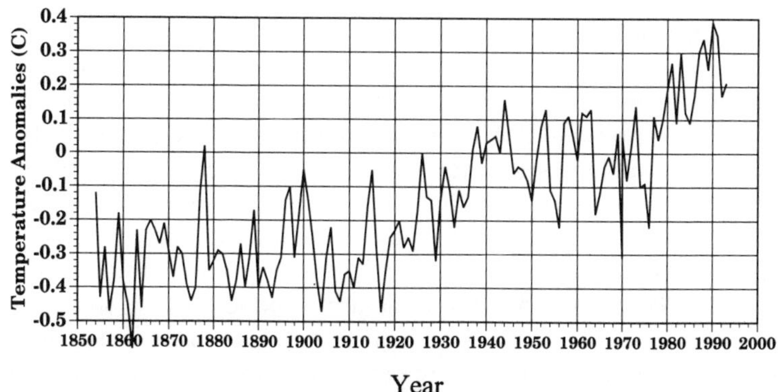

FIGURE 5.3. Variation of the earth's temperature during the past century. These data reflect the variation of the temperature from an arbitrary zero level based on the average temperature during the period 1950–1979. This means that a minus anomaly was cooler than the zero period and a plus anomaly was warmer than the zero period. The data before the year 1900 were based on very few stations and none of the data adequately reflect the temperature of the complete globe. This figure was created from data in P. D. Jones, T. M. L. Wigley, and K. R. Briffa, "Global and Hemispheric Temperature Anomalies—Land and Marine Instrumental Records," in *Trends '93: A Compendium of Data on Global Change*, ORNL/CDIC-65, ed. T. A. Boden, D. P. Kaiser, R. J. Sepanski, and F. W. Stoss (Oak Ridge, TN: Oak Ridge National Laboratory, 1994), pp. 603–608.

determined by ice core measurements, stratification layers of ocean sediment (or other geologic indicators), and tree rings.

The global temperature for the past century is shown in Figure 5.3. The data points on this graph are calculated from weather stations throughout the world. A discussion of these temperature measurements and their meaning constitutes a large part of this chapter, but first let's look at the methods for obtaining global temperature.

SATELLITE MEASUREMENTS

The only technique that offers a truly global approach for earth temperature measurement employs a satellite with a sensor directed toward the earth. This sensor observes a large area beneath the satellite, and as it travels around the earth, the sensor will ultimately sweep over and measure the temperature of the entire globe. It is no small task to ensure that the temperature indicated with the satellite sensor reflects the "true" temperature of the atmosphere near the earth itself. Since the goal is to monitor the air temperature at the earth's surface, the satellite sensors must be able to compensate for the effect of the earth's outer atmosphere.

The satellite-based measurement technique used by space scientists is called *passive microwave radiometry*. In 1978 the first of these satellite sensors, also called "microwave sounding units" (MSUs), was launched aboard the TIROS-N satellite series by the National Oceanic and Atmospheric Administration (NOAA). The MSU senses the microwave emissions from the earth and its atmosphere using four independent measurements; by using computer calibration functions, it can provide the most precise monitoring of global temperature available today. These MSUs are calibrated regularly by measuring the temperature of the cosmic radiation reaching earth from outer space after each global scan. Because cosmic radiation has a constant temperature, it is used as the reference point for calibration.

There are generally at least two satellites in orbit carrying MSUs. The data from each satellite can be used for comparison and cross-checking. Each MSU can measure a 2,000-kilometer path (about 1,080 miles) beneath the satellite's orbit. The measurement differences between one MSU and another are only 0.05°C, based on two-day averages, and 0.01°C, based on monthly averages.[3] This high precision is very important because we are looking for extremely small changes in the earth's temperature.

Unfortunately, since our satellite MSU measurement program is so new, and the data are only available for the period since 1978, it is impossible to draw firm conclusions about long-term trends. The variation of the earth's temperature as measured by satellite is shown in Figure 5.4. These data indicate measurable short-term trends but no longer-term (ten-year) warming trend, a finding that is inconsistent with computer models' predictions of global warm-

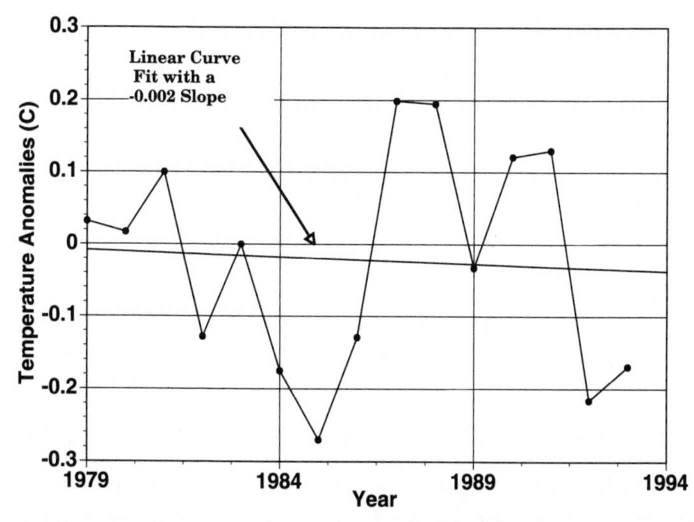

FIGURE 5.4. Satellite measurement of the earth's temperature from 1979 to 1993. The linear trend line for the satellite data actually shows a cooling for the period of measurement. However, this trend is probably not significantly different from a zero trend line, which would indicate no trend at all. The data reflect temperature variations from an arbitrary zero level based on the average temperature during the period 1982–1991. This means that a minus anomaly was cooler than the zero period and a plus anomaly was warmer than the zero period. The data for this figure were taken from R. W. Spencer and J. R. Christy, "Global and Hemispheric Tropospheric and Stratospheric Temperature Anomalies from Satellite Records," in *Trends '93: A Compendium of Data on Global Change*, ORNL/CDIC-65, ed. T. A. Boden, D. P. Kaiser, R. J. Sepanski, and F. W. Stoss (Oak Ridge, TN: Oak Ridge National Laboratory, 1994), pp. 629–634.

ing. It is a shame that the satellites have not been up for the past century; then we would know if the earth really is warming.

EARTH-BASED INSTRUMENT MEASUREMENTS

To determine the global temperature from the earth's surface, meteorologists gather individual temperatures from meteorological stations all over the world and average them. This is not simply a process of adding the temperature values together and dividing by the number of stations. Because these stations are not spaced evenly over the globe, each one represents a different-size area. For example, one might represent 10 square miles and another 1,000 square miles, and these differences must be accounted for. Furthermore, the time of day for each temperature measurement must be addressed to account for temperature variations throughout the day. Today, these temperature records are available from hundreds of sites, but in the past there were far fewer sites. There are many inconsistencies in the temperature measurement process that are indicated in the sections below.

TEMPERATURE MEASUREMENT OVER THE YEARS

First, let us look at the development of the thermometer and see how the accuracy of this tool has varied. Instruments were invented to measure temperature early in the seventeenth century. The earliest ones used the expansion of air with increasing temperature as the sensor mechanism. However, these air-based thermometers were affected by variation in atmospheric pressure, which caused serious errors in the readings.

The first sealed-in-glass thermometers, similar to the ones used today, were invented in about 1650 in Florence, Italy. These thermometers depended on the expansion and contraction of an enclosed liquid with changes in temperature. The first ones used red wine as the indicating liquid, thus the name "spirit-in-glass" thermometers. Over the years mercury has become the most commonly used liquid. During the seventeenth and eighteenth centuries, because centers of scientific learning and research were isolated from each other, the methods of making thermometers varied greatly. All thermometers were completely hand made; there were no mass production procedures used.

On these early thermometers, temperature values were indicated by markings on a paper scale that was positioned behind the glass tube. With time these paper scales would get loose and move about. At the close of the eighteenth century the technique of etching the scale on the glass tube became the practice. This was an essential improvement in that it permitted readings from different thermometers to be compared when the same scales were used.

The problems with scales did not stop here. Some scales were defined to increase with temperature, while others decreased with temperature. These two approaches were upside-down to each other! Also, various researchers used different reference points, which meant that the temperature change associated with one degree varied from scale to scale. In the Utrecht University Museum, there is a thermometer, dated 1754, with eighteen scales, none of which is the common centigrade, or Celsius, scale, although it does contain the Fahrenheit scale, still in use in the United States today.[4] Because of this diversity of scales, most temperature data before the middle of the nineteenth century could not be compared from site to site.

PLACEMENT AND DESIGN OF WEATHER STATIONS IS CRITICAL

As if the issue of temperature scales weren't bad enough, designing, locating, and subsequent changing of measurement sites makes for even worse problems. Measurement inconsistencies occur with local microclimates, such as frost hollows, wide stretches of sandy soil, or exceptionally windswept ridges. Widely divergent temperature averages have been reported from such locations. Many early sites used for temperature observation were inappropriate, and each time a site was changed, there could be a total disruption in the consistency of the measurements. One study reports on sixteen U.S. weather stations where there were ninety-five "important site changes recorded." During the period from 1873 to 1950, all U.S. stations were moved an average of six times each.[5] When meteorologists observe abrupt differences in the data from these site changes, they apply "correction factors;" that is, they adjust the temperature values to make the values compare before and after the move.

Placement of thermometers influences temperature measurement. In the 1700s scientists placed them inside north-facing, unheated rooms, thinking this would be the same as measuring the outside temperature. Others placed thermometers just outside a north-facing window. But buildings heat up, causing the measured temperature to be different from atmospheric temperature.[6]

Over time it was decided that the only way to measure the "true" temperature of the air was to place the thermometer in a protective boxlike container that would shield it from the sun's radiation but would permit the air to flow freely past it. Many designs and concepts were tried before a double-louvered design was generally accepted toward the end of the nineteenth century. This standard weather station, called the Stevenson Screen, is attributed to Thomas Stevenson.[7] Stevenson was the father of the famous nineteenth-century writer Robert Louis Stevenson, who wrote *Treasure Island, Kidnapped,* and *The Strange Case of Dr. Jekyll and Mr. Hyde.*

CHANGING CALIBRATION TECHNIQUES

Many calibration techniques have been used for setting the scale of thermometers. The most common points of reference have been the melting point (or freezing point) of ice and the boiling point of water. Both of these points are dependent on the purity of the water used; impurities in water will lower its freezing point and raise its boiling point. Since the techniques of water purification have evolved through the years and very pure water is a product of this century, these reference points have undoubtedly changed. In addition, the boiling point of water is dependent on barometric pressure and must be corrected for. This means that if the boiling point of water is used as a reference point, the atmospheric pressure must be measured at the same time and a correction factor calculated.

Other reference points that have been used are body heat and the temperature in the cool cellars of observatories. Neither of these points is constant in nature. For instance, after using 98.6°F for years as the standard body temperature, medical researchers have recently found that figure to be inaccurate and to vary from individual to individual. When Daniel Gabriel Fahrenheit, a German instrument maker who lived in Holland, published his famous Fahrenheit scale in the Philosophical Transactions of the Royal Society in 1724, he used 96°, not 98.6°, as the body temperature reference point.[8] This means that the reference points used for standardization of thermometers have not been constant.

In 1776 the Royal Society of England created an ad hoc committee headed by Henry Cavendish to study the state of calibrating thermometers. (Henry Cavendish was a chemist and physicist known for proving that water was not an element and for measuring the force of gravity.) Cavendish's committee found that the thermometers then in use varied by over 3°F. It must be remembered that this study covered thermometer variation only in England and possibly nearby parts of Europe. At this time many countries had their own equivalent of the English Royal Society. These organizations had differing ideas about setting standards, with no international organization to coordinate them. The European scientific community was not in touch with the Middle East or Asia, nor was there appreciable scientific endeavor of any kind south of the equator.

The Cavendish committee recommended both the apparatus and the procedures for setting the scale using the boiling point and freezing point of water. This was probably the first successful attempt at standardizing the thermometric scale. However, these recommendations were not used until 1852 at the Kew Observatory, which was a site for long-term weather measurements in England. It was not until near the end of the nineteenth century that general strides toward true calibration standards were made in national and international laboratories in Europe and America. The Metric Treaty, standardizing calibration procedures, was signed by a few countries in 1875 and is now followed by about fifty countries. The first official temperature scale approved by the treaty (adopted in 1889) was the Normal Hydrogen Scale, which used the boiling (100°C) and freezing points (0°C) of water as calibration points. These standards have been modified several times since and major changes have been instituted in the standard reference points as recently as 1990.[9] Each time the standards are changed, there exists a possibility that corrections to historical data will not be handled consistently.

AVERAGING TECHNIQUES VARY

There are several ways to determine the average, or mean, temperature for any particular day. The consistency of the resulting number used for the average depends not only on the mathematical approach but also on the time each measurement was taken. Two common approaches are as follows:

Method A. Maximum temperature for the day plus minimum temperature for the day divided by two; $(T_{max} + T_{min})/2$.

Method B. The sum of the temperatures taken each hour divided by 24; $(T_1 + T_2 + T_3 + \cdots T_{24})/24$.

If everyone at every measurement station used one or the other of these methods according to a strict protocol, the results would be reliable. Unfortunately, at almost every station, the method of averaging has changed over time, and meteorologists have had to make adjustments of up to ±1°C (±1.8°F) to account for this difference in averaging method. By "adjustment" we mean that the measured data are "corrected" (changed) according to the researchers' assessment of what the numbers should really be! Raymond Bradley, University of Massachusetts, and P. D. Jones, University of East Anglia, have summarized the historical land-based temperature data and discussed the sources of error in detail. They relate that there are only "15 countries that have followed a consistent methodology since the 19th century."[10] Considering that most countries in Europe use different observation times and different methods of calculation,[11] what differences must occur worldwide?

Table 5.1. Official Local Climatological Data from the National Weather Service at Los Angeles International Airport[a]

Hour	Sky cover (tenths)	Temperature (°F)	Relative humidity
		January 4, 1993[b]	
1	0	46	40
4	4	42	49
7	10	45	37
10	10	51	34
13	10	58	22
16	10	54	42
19	10	53	45
22	10	50	54
		April 3, 1993[c]	
1	3	59	75
4	2	56	90
7	0	59	90
10	0	73	52
13	0	68	73
16	1	67	73
19	3	62	73
22	0	62	81

(*continued*)

To give an example of the differences, typical data from the United States Weather Bureau for Los Angeles is reproduced in Table 5.1. It is interesting to note that the observations are reported to only ±1°F (about 0.5°C). If the average temperature for each day in Table 5.1 is calculated by each of the equations discussed above and compared to the average provided by the Weather Bureau, the values in Table 5.2 are obtained. It appears that the government procedure is to use Method A. The difference between Method A

Table 5.1. (continued)

Hour	Sky cover (tenths)	Temperature (°F)	Relative humidity
		July 13, 1993[d]	
1	10	66	84
4	10	66	81
7	10	68	68
10	9	75	52
13	0	73	57
16	1	71	66
19	0	65	78
22	0	64	84
		October 20, 1993[e]	
1	3	62	87
4	3	58	84
7	3	61	56
10	0	81	24
13	0	76	50
16	0	70	76
19	0	68	66
22	0	67	73

[a]Adapted from "Local Climatological Data, Monthly Summary," for January, April, July, and October, Published at the National Climatic Data Center by the U.S. Department of Commerce, 1993. (Observations at three-hour intervals; Latitude 33 56′ N, Longitude 118 24′ W, Elevation 97 feet.

[b]Max. T = 58; Min. T = 41; Average T = 50; January Average T = 56.2

[c]Max. T = 79; Min T = 55; Average T = 67; April Average T = 62.8

[d]Max. T = 78; Min T = 63; Average T = 71; July Average T = 69.5

[e]Max. T = 90; Min T = 56; Average T = 73; October Average T = 68.4

and Method B can result in differences of 3° to 4°. Note that Los Angeles with its very mild climate does not have the extreme diurnal temperature variations or the extreme seasonal temperature variations of latitudes farther north, variations that would amplify these calculation errors in other parts of the world.

Table 5.2. Summary Of Average Temperature Calculations

Date	U.S.W.B. ave. T[a] (F°)	Temp. by Method A[b] (°F)	Temp. by Method B[c] (°F)
1/4/93	50	49.5	50
4/3/93	67	67	63
8/13/93	71	70.5	68
10/20/73	73	73	68

[a]As stated in the U.S. Weather Service Local Climatological Data, Monthly Summary.
[b]Method A is the maximum temperature plus the minimum temperature divided by two.
[c]Method B is the hourly temperature observation divided by 24, using all of the data provided in the U.S. Weather Service Summary.

GROWTH OF CITIES CREATES URBAN HEAT CENTERS

Besides the inconsistencies found in varying sites, the growth of cities brings about artificial temperature changes. As cities grow, much of the cooling greenery is replaced by asphalt and concrete, which act as heat sinks for radiant energy and gradually increase the temperature in the area. A heat sink is any large mass, in this case the streets and buildings, that absorbs heat and changes temperature very slowly as compared to the atmosphere. Hence, a large concrete building will heat up slowly during the day and not completely cool off at night. (The greatest heat sinks on earth are its oceans.) Many weather stations are located within these urban heat centers, and they have provided a record of excessively increasing temperatures.

The presence of the urban heat centers is well documented in both Europe and North America. Locations at Oxford, England; Toronto, Canada; Fairbanks, Alaska; Washington, D.C.; and Phoenix, Arizona have increased in average temperature by up to several degrees in a few short decades.[12] One wonders how much of the currently reported global warming is really due to the placement of weather stations in or close to these urban heat centers!

LIMITED MEASUREMENT SITES

Over the years, there have been scientific organizations that kept temperature records for long periods and developed procedures that permitted the data to be reasonably consistent throughout the years, but these were few and far between. For example, observations were made for over 100 years (about 1750 to 1850) at the Kew Observatory in England. Unfortunately, these measurements were made inside a wooden box attached to the north wall of the observatory, which makes them only approximate actual temperatures. Observations have been made at the Radcliffe Observatory at Oxford, England, since 1815. However, the Oxford site has been influenced by urban heating caused by population growth since about 1960.[13]

The most important flaw in the global temperature concept is the fact that the data do not represent the entire globe. In a landmark publication, "Northern Hemisphere Surface Air Temperature Variations: 1851–1984," a group of highly respected climatologists from England and the United States make the following statement:

> Since the nineteenth century, the station temperature network has expanded considerably. At present most of the land surface of the Northern Hemisphere is adequately covered. However, for periods prior to 1950, significant parts of the hemisphere are not represented. Prior to 1900, . . . the effects of reduced coverage may be substantial. The significance of such changes in coverage has not yet been properly assessed.[14]

So climatologists calculate temperatures for the Northern Hemisphere based on information from the land surface and little or no data from the oceans, which cover most of the hemisphere!

The situation in the Southern Hemisphere is far worse. Figure 5.5 shows the number of "valid" stations over the years for the Northern Hemisphere and the Southern Hemisphere on the same scale. A valid station is one in which the data are consistent or "correctable" to the standard calculation method over the time period in question. In 1900 there were only 45 weather stations with acceptable temperature data in the Southern Hemisphere as compared to 509 in the Northern Hemisphere. There are only

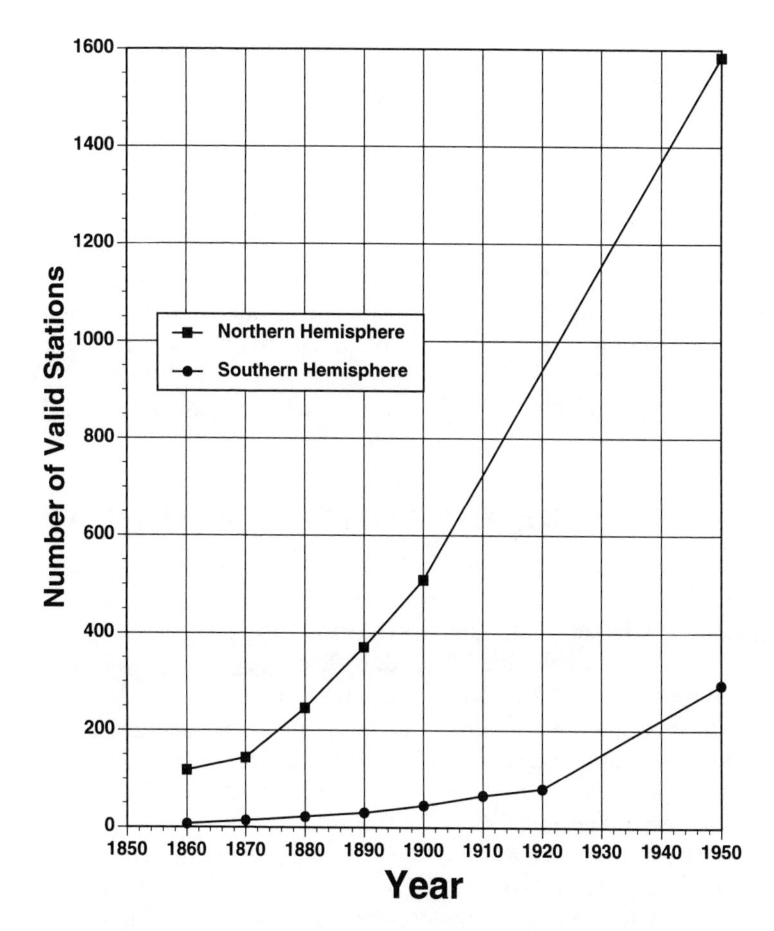

FIGURE 5.5. Number of valid weather measurement stations from 1860 to 1950. The data used in this figure were taken from P. D. Jones et al., "Northern Hemisphere Surface Air Temperature Variations: 1851–1984," *J. Climate Appl. Meterol.* 25 (1986):162–163, for the Northern Hemisphere and P. D. Jones, S. C. B. Raper, and T. M. L. Wigley, "Southern Hemisphere Surface Air Temperature Variations: 1851–1984," *J. Climate Appl. Meterol.* 25 (1986):1213, for the Southern Hemisphere.

thirteen numbers to account for the average temperature in the whole of Africa and another thirteen to account for all of South America! It requires a leap of faith to believe that this skimpy data can be used to represent half of the world's temperature, much less distinguish a change of 0.9°F over a century!

A final point about the coverage of temperature data. There are essentially no data for the polar regions. There was only one valid station above 75° N latitude prior to 1920 (0° N is the equator and 90° N is the North Pole). In the Southern Hemisphere there are virtually no data from 60°–90° S latitude. Data from the Antarctic (South Pole) was nonexistent prior to the 1950s. The information on which our century-long "trend" is based uses slightly over 50% coverage of the land in the Northern Hemisphere and only 29% coverage of the land in the Southern Hemisphere.[15]

THE UNITED STATES, A WELL-MEASURED AREA

The weather data for the United States have been taken at some 1,219 stations since 1900. A group of climatologists from NOAA's National Climatic Data Center in North Carolina, headed by Dr. T. R. Karl, have evaluated these data and provided a well-documented picture of the contiguous United States.[16] The data have been corrected for the "time-of-observation-bias," changes in location of the stations, changes of instrumentation used at the stations, and urban heat center bias. The results of this evaluation are presented in Figure 5.6.

There are essentially no trends observed in the contiguous United States from 1900 to 1990, according to Karl.[17] Although the United States cannot be considered to represent the whole earth or even the Northern Hemisphere, it can be said to be a well-measured system.

In order to give a clear perspective of the annual variation within which climatologists are trying to determine a change of less than a degree during the past century, the temperatures for the Los Angeles area are presented in Figure 5.7 for the year 1993. The year 1993 was not remarkable for temperature extremes; it can be con-

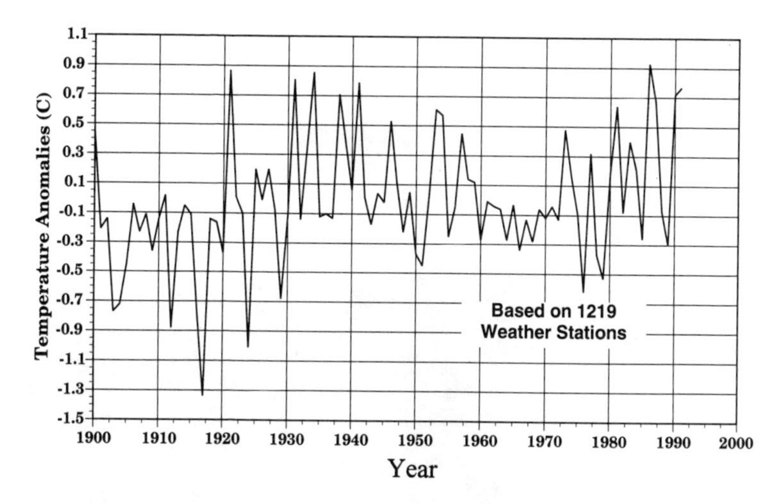

FIGURE 5.6. The variation of temperature in the contiguous United States from 1900 to 1991. During the period 1948–1987, these data show a cooling trend of –0.06°C (–0.11°F). There is clearly a warming trend before 1950, however. The data reflect the variation of the temperature from an arbitrary zero level based on the average temperature during the period 1961–1990. This means that a minus anomaly was cooler than the zero period and a plus anomaly was warmer than the zero period. The data used in this figure were taken from T. R. Karl, D. R. Easterling, R. W. Knight, and P. Y. Hughes, "U.S. National and Regional Temperature Anomalies," in *Trends '93: A Compendium of Data on Global Change*, ORNL/CDIAC-65, ed. T. A. Boden, D. P. Kaiser, R. J. Sepanski, and F. W. Stoss (Oak Ridge, TN: Oak Ridge National Laboratory, 1994), pp. 686–736.

sidered a typical year. Looking for a change of a small part of a degree during the past century in the noise of the real changes from year to year continues to stretch the imagination!

OCEAN MEASUREMENTS VERSUS LAND-BASED MEASUREMENTS

Our discussions so far have focused on land-based temperature measurements. The oceans cover over 70% of the earth's surface.

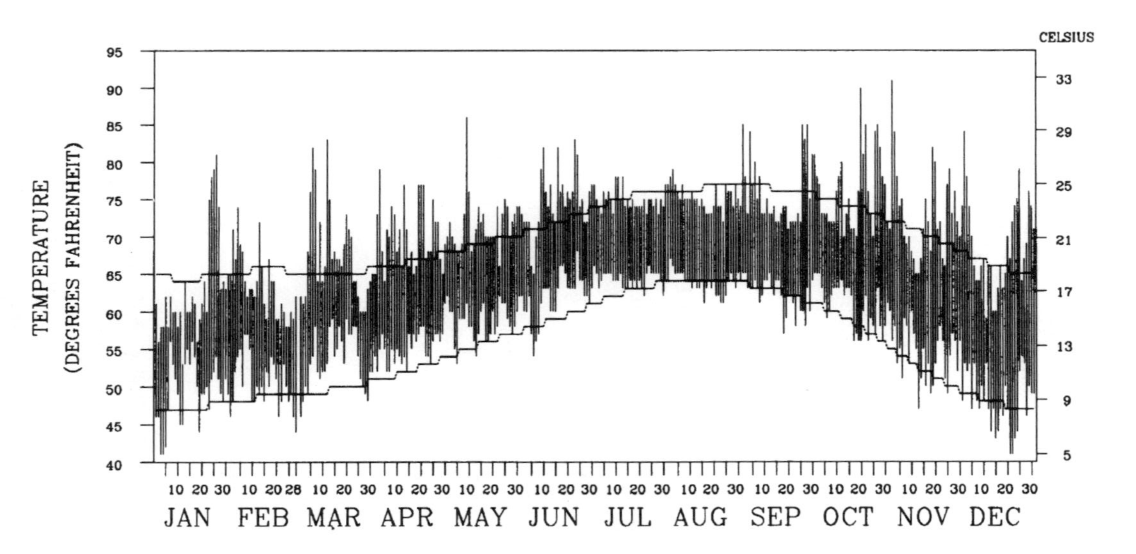

FIGURE 5.7. Temperature variation for Los Angeles International Airport for 1993. Note that even in the mild climate of Los Angeles the day to night and seasonal temperatures vary by 40° to 60°F. Keep this in mind when the search is for a half of a degree per century. This figure was drawn from the Official Weather Data from the United States Department of Commerce, National Climatic Data Center in Asheville, North Carolina.

The oceans have a very complex current–temperature relationship beneath the surface that affects the temperature at the surface of the water and hence the air temperature above the surface. Of course, it is this air temperature that is essential when trying to determine the temperature of the earth's atmosphere.

Historically, weather measurements have been made on board ships traveling all over the world. Ocean travel was extremely limited before the sixteenth century but has grown dramatically in the twentieth century. Currently, there is a Comprehensive Ocean-Atmosphere Data Set (COADS), which includes information on eight weather variables, including air and sea-surface temperature, wind speed and direction, barometric pressure, humidity, and cloudiness. This data set is the result of a cooperative project to collect global weather observations taken near the ocean's surface since 1854, according to the description provided by NOAA scientists headed by Scott Woodruff at the Environmental Research Laboratories in Boulder, Colorado.[18] Historically, land-based and ocean temperature data have been assembled and evaluated by different organizations.

COADS suffers from many of the same problems that plague the land-based data. The early coverage was very limited and only in the past two to three decades has the percentage of coverage of the globe been close to adequate. The methods of obtaining temperatures over the ocean have changed and are not consistent; standards and calibration methods used by different countries have been different. Older records often did not indicate the information needed to evaluate the method used. As a consequence, a major screening program has been necessary to refine these data and make them as accurate as possible.

For example, there have been several techniques employed for obtaining the surface temperature of the oceans. During some years, wooden buckets were dropped into the ocean, then pulled to the deck of the ship where the temperature of the water was measured. During other, but certainly overlapping periods, canvas buckets were used, which caused a temperature variation due to evaporative cooling from the canvas. At still other times, the temperature was taken by measuring the intake water used to cool the ship's engines. Unfortunately, these different techniques give varying results. Even more unfortunate is the fact that the details of the method were often not indicated in the ship's logbooks.

These problems pale, however, when a side-by-side comparison is made between land-based temperature data and the sea-based air temperature data. These two data sets do not correlate at all! This means that the trends observed on land do not agree with the trends observed at sea! How is it possible for scientists to state that there is a global warming trend when the major sources of temperature information do not even agree?

NATURAL CLIMATE VARIATION AND THE LITTLE ICE AGE

During the last 500 years, a significant cooling took place in Europe and North America that was so severe it has been called the Little Ice Age. It has been thought that this was a global cooling event; however, Raymond S. Bradley at the University of Massachusetts and Philip D. Jones at the University of East Anglia, England, have edited a major work, *Climate Since A.D. 1500,*[19] in which this conclusion is questioned.

It is interesting to look at the information developed by climatologists concerning the Little Ice Age. It clearly had a major impact on Western Europe and North America, but the impact on Russia was almost nonexistent, and the impact on Iceland and Australia seemed to have a very different time frame. These differences have implications for our perception of "global temperature." The scientific center of the Western world is in Western Europe and North America. Is it possible that the Little Ice Age was simply a regional event that affected those areas where the majority of the data was collected? Can we infer a global event from this limited data base? Let's look at the evidence for the Little Ice Age; it occurred before we had substantial instrumental measurements of the weather.

With very few exceptions our knowledge of the global temperature before the middle of the last century comes from indirect sources such as:

- Historical documents, including ancient inscriptions, annals, chronicles, government records, private records, commercial records, personal papers and diaries, scientific or quasi-scientific writings, and fragmented early instrumental measurements
- Tree ring measurements (dendrochronology)
- Ice core measurements

Of course, each of these techniques provides a time series of information at a single location on the earth. Further, such techniques as tree ring and ice core measurements provide only seasonal information. Tree rings grow only during the tree's growing season. The size of each tree ring is a function of several variables, including temperature. The ice grows or is added to the ice pack only during the snow seasons. To build an accurate knowledge of the global picture requires a global data base.

Bradley and Jones summarize the current understanding about the Little Ice Age as follows:

- The last 500 years have not experienced a monotonously cold Little Ice Age: certain intervals have been colder than others.
- The coldest periods in one region are often not coincidental with those in other regions. There is a geographical variability in climatic anomalies.
- Different seasons may show different anomaly patterns over time.[20]

This means that it was coldest in one part of the world at a different time and for a different period of time than in another. In addition, Bradley and Jones concluded that the decade-to-decade variation was often more dramatic than any longer-term variation.

Parts of the earth appear to be warming from this cooling event. Is it possible that the current warming, if indeed there is a current warming, is due to the earth's natural recovery from this Little Ice Age?—that is, if indeed there was a Little Ice Age on a global scale!

GLOBAL TEMPERATURE ASSESSMENT

All of these inconsistencies result in real uncertainty as to the meaning of the earth's "global temperature" and in whether there is a measurable trend in the global temperature change.

- Only a handful of long-term, consistent records is available. These records are from Europe and the northeastern United States and Canada, and none is truly without the inconsistencies mentioned above.
- These long-term data sets represent only a portion of the Northern Hemisphere.
- These data do not correlate with the ocean-based data.
- When looked at on a regional basis, there are often no trends at all, and sometimes there are cooling trends. For instance, data for the United States indicates a national cooling of 0.06°C (0.11°F) during this century.[21]

Nonetheless, this information has been used by the IPCC, the National Academy of Sciences, and other government panels to conclude that the temperature of the earth has increased during the past century by 0.3–0.6°C.

Can this conclusion be called "good science"? The answer is no!

There are many known problems about the historical climatic condition of our planet that reinforce the uncertainty about a global temperature increase:

- Regional variation in warming and cooling trends.
- Major time differences from region to region for the beginning of warming and cooling trends.
- Lack of understanding of "natural" warming and cooling events such as the El Niño-Southern Oscillation and major volcanic eruptions.
- Lack of consistent temperature data or global measurement for most of the past century.

It appears that climatologists now have the capability in hand to measure the earth's temperature; however, our knowledge of the temperature a century ago is inadequate. The IPCC assessment should indicate at least twice as much uncertainty in the global temperature increase and probably more. Remember the uncertainty that was cited by the IPCC was stated to be within the range of natural variation of the earth's temperature. The temperature values on which this warming number is based are insufficient to make any valid claim of warming over the past century.

REFERENCES

1. W. E. Knowles Middleton, *A History of the Thermometer* (Baltimore: Johns Hopkins Press, 1966), p. 9.
2. Melvin A. Benarde, *Global Warning . . . Global Warming* (New York: John Wiley & Sons, 1992), pp. 15–35.
3. Roy W. Spencer and John R. Christy, "Precise Monitoring of Global Temperature Trends from Satellites," *Science* 247 (1990):1558–1562.
4. Middleton, 65–66.
5. P. D. Jones, S. C. B. Raper, R. S. Bradley, H. F. Diaz, P. M. Kelly, and T. M. L. Wigley, "Northern Hemisphere Surface Air Temperature Variations: 1851–1984," *J. Climate Appl. Meteorol.* 25 (1986):162–163.
6. Gordon Manley, "Central England Temperatures: Monthly Means 1659 to 1973," *Q. J. R. Meterol. Soc.* 100 (1974):391.
7. Middleton, 223–225.
8. Middleton, 75.
9. J. V. Nicholas and D. R. White, *Traceable Temperatures, an Introduction to Temperature Measurement and Calibration* (New York: John Wiley & Sons, 1994), pp. 16–17.
10. Raymond S. Bradley and P. D. Jones, "Data Bases for Isolating the Effects of the Increasing Carbon Dioxide Concentration," in *Detecting the Climatic Effects of Increasing Carbon Dioxide,* DOE/ER-0235, ed. Michael C. MacCracken and Frederick M. Luther (Washington, DC: U.S. Department of Energy, 1985), p. 34.
11. Jones et al., 162–163.
12. Manley, 392; Jones et al., 163–164.
13. Manley, 389–405.
14. Jones et al., 161.
15. Jones et al., 171; P. D. Jones, S. C. B. Raper, and T. M. L. Wigley, "Southern Hemisphere Surface Air Temperature Variations: 1851–1984," *J. Climate Appl. Meteorol.* 25 (1986):1213.

16. T. R. Karl, D. R. Easterling, R. W. Knight, and P. Y. Hughes, "U.S. National and Regional Temperature Anomalies," in *Trends '93: A Compendium of Data on Global Change*, ORNL/CDIAC-65, ed. T. A. Boden, D. P. Kaiser, R. J. Sepanski, and F. W. Stoss (Oak Ridge, TN: Oak Ridge National Laboratory, 1994), pp. 686–736.
17. Ibid., 688.
18. Scott D. Woodruff, Ralph J. Slutz, Roy L. Jenne, and Peter M. Steurer, "A Comprehensive Ocean-Atmosphere Data Set," *Bull. Am. Meterol. Soc.* 68 (1987):1239.
19. Ramond S. Bradley and Philip D. Jones, eds., *Climate Since A.D. 1500* (London: Routledge, 1992).
20. Bradley and Jones, 659.
21. Karl et al., 688.

PART III

Factors Affecting Climate and Climate Computer Models

SIX

Greenhouse Gases

Let us imagine a world without greenhouse gases in the atmosphere. It would be cold and frozen at –15°C (+5°F), mostly covered by ice pack. The clouds, if there were any, would be made of wisps of aerosol consisting of carbon dioxide and sulfates from the volcanic activity of the ever-changing and evolving globe, and there would be little moisture in the atmosphere. Any life would have evolved under the ice cap covering the oceans. The photosynthesis that provides the nutrients for nearly all plant life could not exist because the ice would block the sun's energy. Perhaps you have seen pictures of the strange aquatic plants and animals that have evolved in the dark bottom of the deep oceans. These weird creatures would probably represent the most advanced life on earth, for land-dwelling animals, including humans, could not exist.

Enough of this speculation. Thank goodness that the planet has an abundance of greenhouse gases, that green plants grow, and that humankind has the ability to adapt to extreme climates.

Greenhouse gases have existed in our atmosphere in varying amounts for much of the earth's existence—billions of years. In order of their probable effect, the greenhouse gases are: water vapor, carbon dioxide, methane, man-made chlorofluorocarbons (CFCs), and other trace gases. The sum of all these gases in the atmosphere, with the exception of water vapor, amounts to a few hundred parts per million (ppm). Water vapor varies from a percent or two (1% is equal to 10,000 ppm) in very humid and wet areas to very low levels (a few hundred ppm) in dry, desert climates. A ppm is about one drop in 10 gallons of water, or about one inch in 16 miles!

Water vapor is by far the most important greenhouse gas, but it has received almost no attention from global warming theorists. This lack of attention is probably due to the fact that no one gets excited about the evils of water vapor. Water is too prevalent on the earth. It is too much a part of our everyday lives. Even though it is one of the combustion products of fossil fuels, along with carbon dioxide, it hardly seems threatening. In truth, it is far more important than all the other greenhouse gases in moderating our climate.

Climate scientists do not understand the extremely complex interactions between water vapor, clouds, oceans, and ice packs that make up the hydrological cycle, and their computer climate models do not adequately account for them. These various forms of water cause different and often opposite effects when introduced into climate models. The hydrological cycle is so important in understanding our climate that it warrants a complete chapter (Chapter 8).

CARBON DIOXIDE

Because carbon dioxide (CO_2) is the most commonly publicized gas in global warming, let us look first at its role in the greenhouse effect. Carbon dioxide and the carbon cycle are nearly as complex as the hydrological cycle. Carbon exists in various forms in plants, animals, soils, rocks, and the oceans, in addition to the atmosphere. However, scientists do not know where almost half of the carbon dioxide that is emitted into the atmosphere ends up!

On the one hand, all animals use oxygen from the air to fuel their life functions, and with every breath they emit carbon dioxide gas into the atmosphere. Every fire that burns from wood, coal, natural gas, biomass, or petroleum produces carbon dioxide and water vapor as major combustion products. On the other hand, all plants absorb carbon dioxide from the atmosphere, utilizing it along with water and sunlight to produce the food they need to grow—the process known as photosynthesis. We must never forget that carbon dioxide is essential for human life on earth and that it cannot be simply eliminated from the world.

The cycle of carbon on the earth is illustrated in Figure 6.1. As you can see from this diagram, carbon dioxide is removed from the atmosphere not only by plants (biota), but also by oceans and lakes, because carbon dioxide is soluble in water. (A body that removes carbon dioxide from the atmosphere and stores it is referred to as a carbon "sink.") When carbon dioxide dissolves in water, it forms carbonic acid; this is the form of carbon dioxide that you drink in most "carbonated" soft drinks. Some of the carbon dioxide in oceans and lakes is reemitted into the atmosphere, but most of it is either used in photosynthesis by ocean plants or, by forming compounds with other chemicals in the water, deposited as carbonate sediment on the ocean or lake bottom.

Scientists still do not fully understand the carbon cycle. They have a good idea how much carbon dioxide is emitted each year by means of combustion from all sources, but they do not know where all of it ends up. There is a major imbalance between the amount emitted and the amount measured in the air. This imbalance, which adds up to some 30–40% of carbon dioxide emissions, is sometimes referred to as the "mystery of the missing carbon." Scientists disagree on the relative amounts dissolved in the oceans, absorbed by plants and trees, and deposited in the organic components of the soils. Studies are published regularly that point out overlooked sinks for carbon, such as deep rooted grasses,[1] forests,[2] or imbalances measured in the oceans.[3] Scientists around the world are trying to find the answers to this perplexing question.

The government has taken an active interest in solving this problem of the missing carbon dioxide. In 1985, the U.S. Department of Energy published a series of state-of-the-art reports on various aspects of the carbon dioxide issue. These reports were peer reviewed by a committee of the American Association for the Advancement of Science to insure their completeness and accuracy. Of particular interest here is the report edited by John R. Trabalka, of Oak Ridge National Laboratory, entitled *Atmospheric Carbon*

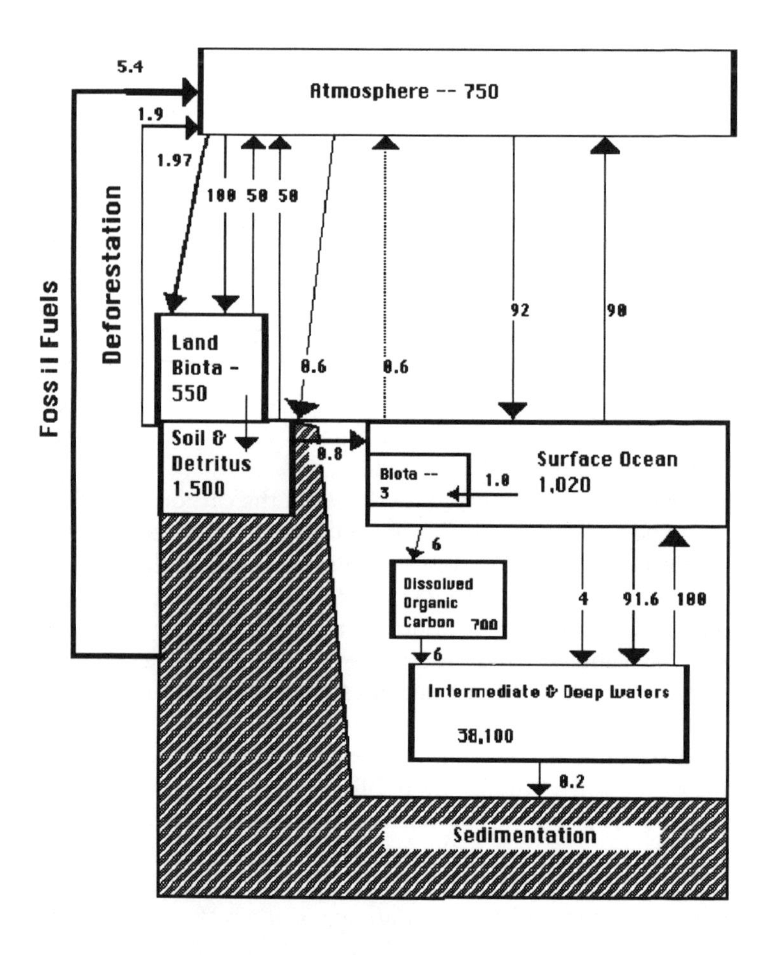

FIGURE 6.1. The carbon cycle (1980–1990). Reservoir numbers represent gigatons of carbon and the flux numbers are gigatons per year. Adapted from data in U. Siegenthaler and J. L. Sarmiento, "Atmospheric Carbon Dioxide and the Ocean," *Nature* 365 (1993):119–127.

Dioxide and the Global Carbon Cycle. Two very important conclusions are expressed in this work. The first is that only about 60% of carbon dioxide emissions into the atmosphere can be accounted for by current understanding of sinks; this means that scientists do not know where 40% of carbon emissions end up. The second is that, without this understanding, scientists cannot hope to quantify adequately the role of carbon dioxide in computer climate models.[4] In other words, climate models are being run that include huge uncertainties with respect to the carbon cycle, and yet the results are being accepted as valid.

More recently, professors U. Siegenthaler of the University of Bern, Switzerland, and J. L. Sarmiento of Princeton University reviewed the current understanding of the carbon cycle in a 1993 feature article in *Nature*.[5] They report that further studies, to some degree, have improved the knowledge regarding the imbalance in carbon dioxide sinks relative to emissions. These scientists think they know where about 70% of the carbon is instead of 60%, but that still leaves 30% of carbon dioxide emissions unaccounted for.

How can climate scientists create computer models of the carbon cycle if they don't fully know the processes by which carbon dioxide is removed from the atmosphere? This lack of understanding on the part of the scientific community has not deterred some scientists in their conviction that global warming is caused by increasing concentrations of greenhouse gases in our atmosphere; their sweeping conclusions are published and used as the basis for drastic policy decisions, as we noted in Chapter 3.

SOURCES OF CARBON DIOXIDE EMISSIONS

There are many natural sources of carbon dioxide, including nearly all living species of animals, microbes in the soils, decaying plants and animals after they die, and gaseous emissions from lakes and oceans. The anthropogenic sources are mainly derived from the burning of fossil fuels and deforestation. The various natural and human-generated sources are summarized, along with the magnitudes of each, in Table 6.1. One of the first things that strikes

Table 6.1. Sources of Carbon Dioxide[a]

Natural Sources	Value (Gt/Yr)[b]	Percent (%)
Oceans	90	45.6
Land biota[c]	50	25.3
Soil and detritus	50	25.3
Subtotal	190	96.3
Anthropogenic sources		
Burning fossil fuels	5.4	2.7
Deforestation	1.9	1.0
Subtotal	7.3	3.7
Total	197.3	100.0

[a]Adapted from data presented in the review article: U. Siegenthaler and J. L. Sarmiento, "Atmospheric carbon dioxide and the ocean," *Science* 365 (1993): pp. 119–127.

[b]A Gt/Yr is the number of gigatons of carbon dioxide that are emitted per year. A gigaton is a billion tons (1 Gt = 1,000,000,000 tons).

[c]Land Biota represents emissions from all plantlife on the earth.

you in Table 6.1 is that, compared to the carbon dioxide that is emitted into the atmosphere from natural sources, the anthropogenic sources are very small, only about 3.7% of the total.

A second and somewhat startling point is that there is essentially no information on the amount of carbon dioxide directly emitted into the air by the respiration of humanity and the animal population. For example, this source is not mentioned by Siegenthaler and Sarmiento; nor is it mentioned in either of the IPCC reports. A conservative calculation would add a quantity at least equal to that from deforestation. If carbon dioxide from human and animal respiration were included in the anthropogenic sources, the amount of carbon unaccounted for in the atmosphere would be substantially increased!

Besides the carbon dioxide that is emitted by the sources indicated in Table 6.1, there are gigantic emissions of carbon dioxide from volcanic eruptions. The carbon dioxide and other gaseous emissions from volcanic activity have been ignored by most climate modelers; these emissions and their potential effect on the atmosphere are discussed further in Chapter 9.

There is no question that the activities of the human race have provided a small addition to the natural carbon dioxide in the atmosphere, both directly and indirectly, and as the population grows by leaps and bounds, this source of carbon dioxide will become more significant.

The direct burning of fossil fuels is well documented. Most governments directly control or regulate the generation of power within their countries; as a consequence, records are reasonably good with respect to the amount of fuel consumed each year and thus the amount of carbon dioxide emissions. These numbers can be provided with much greater confidence than numbers from most other sources. These data are presented in Figure 6.2. Note that the trends for liquid fuels (gasoline and fuel oil) seem to be leveling off, while the trends for solid fuels (coal) and gaseous fuels (natural gas, propane, and butane) continue to rise. The secondary air pollution problems are significant with solid fuels, whereas gaseous fuels are the cleanest burning.

There are major uncertainties associated with all of the numbers in Figure 6.1 and Table 6.1. It is impossible to determine exactly how much carbon dioxide is emitted from the oceans and how much is absorbed by the forests of the world; these numbers represent "best" estimates. A great deal of scientific effort is being utilized to determine these levels with greater accuracy, and in doing so scientists are learning more about the carbon cycle. The fact remains, however, that the numerical data being input into computer models are woefully inadequate and fraught with uncertainty. There is a lot to learn about the carbon cycle and especially the "missing carbon."

CARBON DIOXIDE AND THE ICE AGES

The concentration of carbon dioxide in the atmosphere has varied through the ages. It has changed with the glaciation cycles

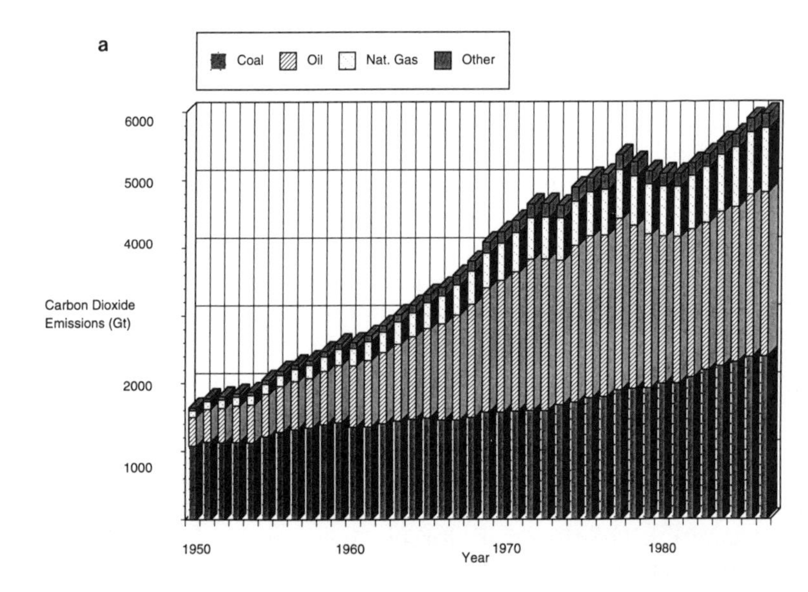
a
Coal
Oil
Nat. Gas
Other
6000
5000
4000
Carbon Dioxide
Emissions (Gt)
2000
1000
1950
1960
Year
1970
1980

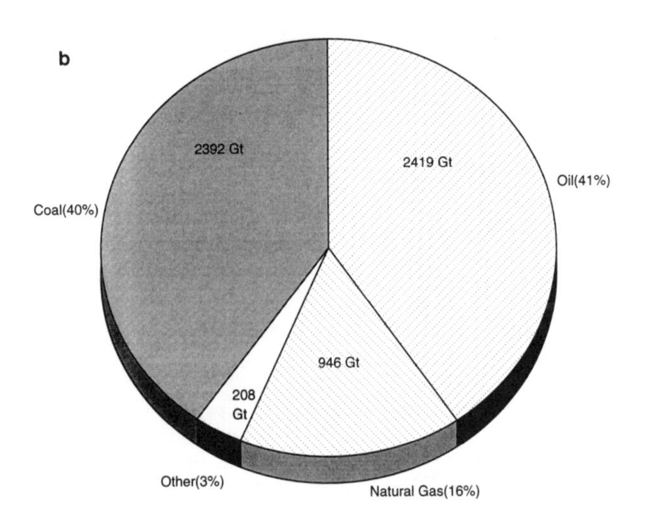
b
2392 Gt
2419 Gt
Oil(41%)
Coal(40%)
946 Gt
208
Gt
Other(3%)
Natural Gas(16%)

(ice ages), being higher during interglacial periods (the warmer periods between ice ages), and lower during times of major glacial advances. Because of this relationship, the question naturally arises—does the carbon dioxide concentration in the atmosphere regulate the earth's temperature or does the earth's temperature determine the carbon dioxide concentration in the atmosphere?

Advocates of the global warming theory argue that atmospheric carbon dioxide concentration regulates the earth's temperature, but other scientists disagree. One scientist who has done a great deal of research on carbon dioxide is Sherwood B. Idso, of the United States Water Conservation Laboratory in Phoenix, Arizona. Idso points out that if the temperature and carbon dioxide historical records are studied carefully,

> changes in atmospheric carbon dioxide concentration never precede changes in air temperature, when going from glacial to interglacial conditions, and when going from interglacial to glacial conditions, the change in CO_2 concentration actually lags the change in air temperature.[6]

Clearly, if decreases in carbon dioxide in the atmosphere caused the ice ages, these decreases should come before the temperature decrease, not after. The fact that atmospheric carbon dioxide changes lagged behind temperature changes suggests that the earth's temperature determines the amount of carbon dioxide in the atmosphere and not the other way around.

Some scientists believe that these periods of glaciation were caused mainly by variations in the earth's orbit with respect to the sun, as was pointed out in Chapter 5, with some possible amplification of the temperature trends due to a change in carbon dioxide concentration. W. J. Burroughs, research climatologist and science

---

FIGURE 6.2. Global carbon emissions from the burning of fossil fuels. a. Trend from 1950 to 1990 (numbers are in gigatons). *Oil* includes fuel oil, diesel, and gasoline. *Other* includes cement manufacturing and gas flaring. b. Percentages from various types of fossil fuels in 1989. Adapted from data in *Trends '91: A Compendium of Data on Global Change*, ORNL/CDIAC-46, T. A. Boden, R. J. Sepanski, and F. W. Stoss, eds. (Oak Ridge, TN: Oak Ridge National Laboratory, 1991), pp. 386–389.

writer, states in his recent treatise *Weather Cycles* that gravitational interactions of the earth with the moon and other planets have created cyclical variations in the earth's solar orbit, with periods that coincide with glacial cycles.[7] As noted in Chapter 5, these orbital variations were first calculated by Milutin Milankovitch, a Serbian engineer, and they have subsequently been observed and well documented. Orbital variations are consistent with the timing of the ice ages and there can be no doubt that if the earth is farther from the sun, it will receive less of the sun's energy and vice versa.

HISTORICAL ATMOSPHERIC CARBON DIOXIDE

Interest in carbon dioxide as a possible cause of climate regulation increased by the mid-twentieth century, but there were few actual measurements and no long-term monitoring programs before that time. Scientists did not begin to monitor the actual concentration of carbon dioxide in the atmosphere until the late 1950s. The first serious effort was initiated in 1958 by Dr. C. D. Keeling of the Scripps Institute of Oceanography, La Jolla, California. Keeling has made continuous measurements at a site at Mauna Loa Observatory, Hawaii, since that time.[8] Figure 6.3, which includes Keeling's data from Mauna Loa, shows the carbon dioxide concentrations for several different time spans. Keeling's data (Figure 6.3a) clearly illustrate the variation between the seasons and the increasing carbon dioxide trend since his measurements began. Note, however, that the upward trend of carbon dioxide in the atmosphere has begun to level out during the last few years covered by this figure. Total emissions of carbon dioxide (shown in Figure 6.2a) show a leveling off in the 1970s and a steady increase throughout most of the 1980s. There is no clear correlation between CO_2 emission levels and the amount of CO_2 measured in the atmosphere. Where does all the carbon end up? The scientists simply do not know.

The concentration of atmospheric carbon dioxide before the industrial revolution has been determined by measuring the atmospheric gases trapped in the ice. To do this, researchers cut a core

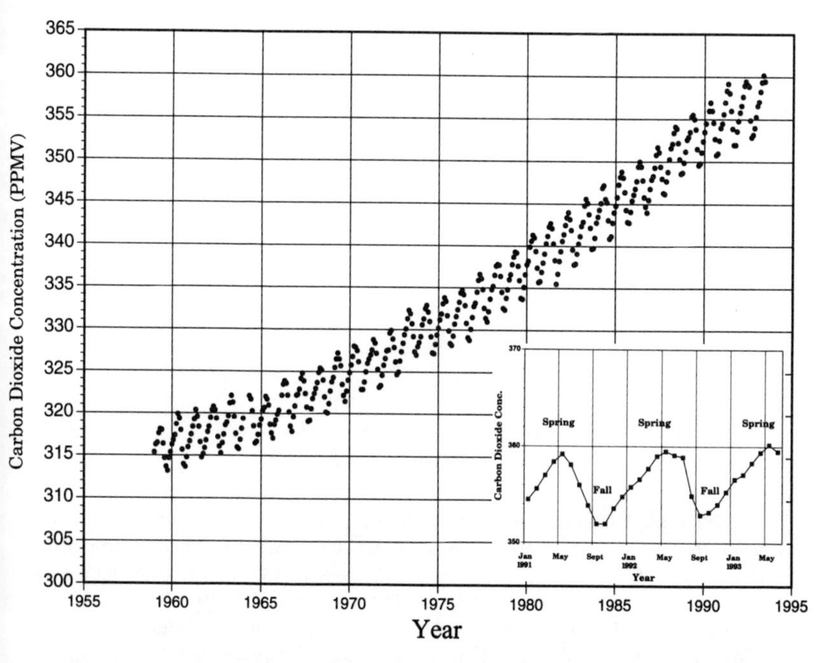

FIGURE 6.3. Carbon dioxide concentration in the atmosphere. a. Values measured directly in the atmosphere at Mauna Loa Observatory, Hawaii, 1958–1993. Data taken from C. D. Keeling and T. P. Whorf, "Atmospheric CO_2 Records from Sites in the SIO Air Sampling Network," in *Trends '93: A Compendium of Data on Global Change*, ORNL/CDIAC-65, ed. T. A. Boden, D.P. Kaiser, R. J. Sepanski, and F. W. Stoss (Oak Ridge, TN: Carbon Dioxide Information Analysis Center, Oak Ridge National Laboratory, 1994), pp. 16–26.

from the ice pack of a glacier and release the trapped gases into instruments that measure the quantities of the various gases. The age of these gases can be determined by the stratification caused by seasonal accumulation of snow and ice. Each year snow and ice are deposited during the winter; during the summer this precipitation stops and some melting may even occur. This disruption in deposited ice causes an observable stratification. And whereas

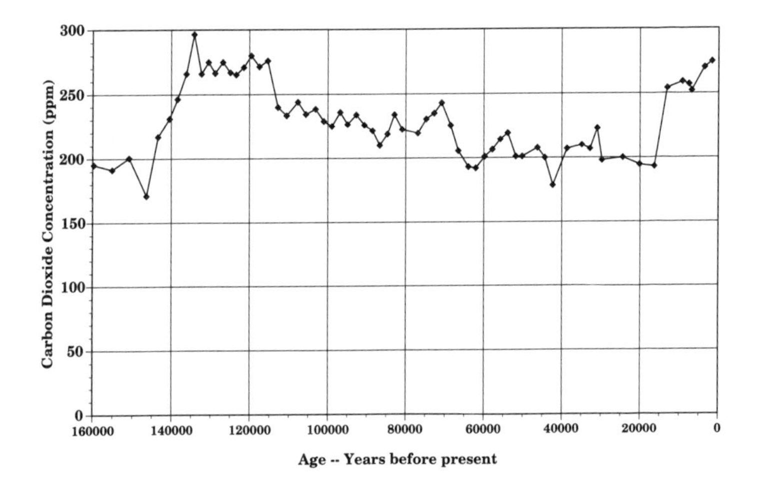

FIGURE 6.3b. Values determined from ice core measurements at Vostok, Antarctica, from 160,000 to 1,700 years before present. Data taken from J. M. Barnola, D. Raymond, C. Lorius, and Y. S. Korotkevich, "Historical CO_2 Record from the Vostok Ice Core," in *Trends '93: A Compendium of Data on Global Change*, ORNL/CDIAC-65, ed. T. A. Boden, D. P. Kaiser, R. J. Sepanski, and F. W. Stoss (Oak Ridge, TN: Carbon Dioxide Information Analysis Center, Oak Ridge National Laboratory, 1994), pp. 7–10.

there are uncertainties associated with this procedure, it provides us with an approximate knowledge of historic levels of carbon dioxide in the atmosphere and the increases and decreases throughout the ages.

The polar ice core at Vostok, Antarctica, provided data showing that the carbon dioxide baseline has varied between about 200 and 300 ppm during the past 160,000 years (Figure 6.3b).[9] Another ice core, taken at Siple Station, Antarctica, provides more recent information—from about 1865 to the present. In 1865 the premodern background concentration of carbon dioxide was about 280 ppm (Figure 6.3c).[10]

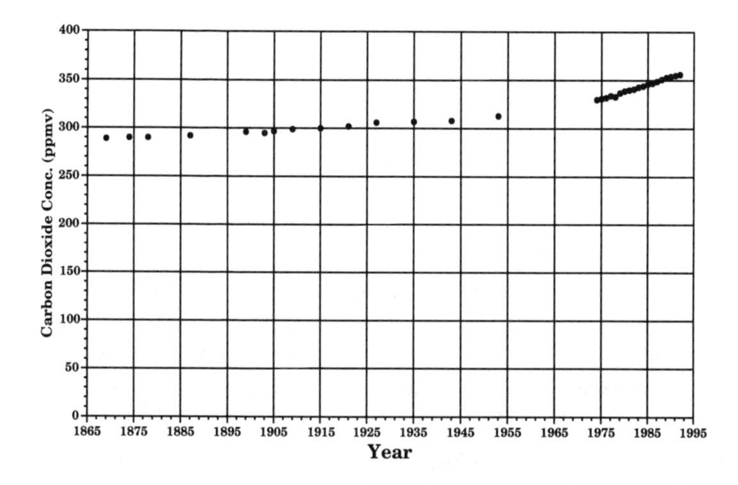

FIGURE 6.3c. Values determined from ice core measurements at Siple Station, Antarctica, 1865–1953, and Keeling's Data from 1958–1003. Data taken from A. Neftel et al., "Historical CO_2 Record from the Siple Station Ice Core," in *Trends '93: A Compendium of Data on Global Change*, ORNL/CDIAC-65, ed. T. A. Boden, D. P. Kaiser, R. J. Sepanski, and F. W. Stoss (Oak Ridge, TN: Carbon Dioxide Information Analysis Center, Oak Ridge National Laboratory, 1994), pp. 11–14.

Older ice core and ocean sediment core data provide information going back several million years. These older data show that the carbon dioxide concentration has been both higher and lower than that determined by more recent measurements. According to the DOE report *Atmospheric Carbon Dioxide and the Global Carbon Cycle* (mentioned before), when the earth was very young, the concentration of carbon dioxide was thought to be on the order of 0.4% (4,000 ppm) or higher,[11] more than ten times today's level! This very high carbon dioxide concentration would be associated with an earth warm enough to permit the beginnings of life.

BENEFITS OF INCREASED CO_2 IN THE ATMOSPHERE?

There is no doubt that the concentration of atmospheric carbon dioxide is increasing from the level of the 1800s. But is this increase a sign of impending doom, or is it a blessing, or does the truth lie somewhere in between? Will the potential increase in temperature result in turmoil or a better quality of life?

In Chapter 1 we discussed Svante Arrhenius, the nineteenth-century scientist who was one of the first to recognize carbon dioxide as a greenhouse gas. Many global warming doomsayers point out that Arrhenius predicted global warming as a result of increases in carbon dioxide, but some of them go on to say that Arrhenius warned of the dangers of these increased levels. For example, Harold W. Bernard, Jr., an atmospheric scientist, states in his book *Global Warming Unchecked* that Arrhenius gave the "very first warning." He goes on to say that "Arrhenius was far ahead of his time, and his dire forecast was met with scorn by the scientific community."[12] But Barnard neglects to mention that Arrhenius felt that this warming was good for humankind! Living in Sweden during the last part of the Little Ice Age, Arrhenius felt that the earth was too cold and that the warming and fertilizing effects of carbon dioxide would be beneficial.

It is irrefutable that an increased level of carbon dioxide will benefit the growth of plant life, which is essential for human survival. Humans require plants for sustenance; the cattle, sheep, pigs, and poultry that we raise for food require plants for sustenance as well. Sherwood Idso is enthusiastic about the beneficial effects of increased carbon dioxide. He points out in his well-documented book, *Carbon Dioxide and Global Change: Earth in Transition*, that research with all types of plants has been performed over the past hundred years and has demonstrated that increased levels of CO_2 produce larger plants with taller branches and more extensive root systems. Quantity and size of flowers and fruit are also enhanced. In crops grown with greater levels of carbon dioxide, crop yields are boosted substantially (often over 30%).[13] Agriculture will certainly benefit from higher levels of carbon dioxide in the atmosphere.

Another scientist, B. A. Kimball, of the United States Water Conservation Laboratory in Phoenix, Arizona, has compiled the results of hundreds of studies that have been done on the effects of increased levels of carbon dioxide on plants. He found improved yields with no negative results on plant growth or other problems.[14] Kimball's summary covers a major body of scientific agricultural study of many different crops in many geographic areas. This is an important statement as to the value of carbon dioxide for plant life in general.

A great deal of information on the effect of increased carbon dioxide levels on major crops has been summarized by Jennifer Cure of Duke University in a report prepared for the Department of Energy titled *Direct Effects of Increasing Carbon Dioxide on Vegetation.*[15] Table 6.2 provides data for several of the major farm crops grown around the world, including wheat, rice, corn, and potatoes. Note that total biomass (weight of the plants) increases are generally well over 30%, and crop yields are also up by over 30%. This table shows results with some of the most important crops in the world today, and these findings are very positive endorsements for the fertilizing effects of carbon dioxide.

An experimental study that explored the effects of various factors, including carbon dioxide levels, on plant growth was performed by B. G. Bugbee and F. B. Salisbury, scientists in the Plant Science Department at Utah State University. They determined that the ideal atmospheric carbon dioxide concentration for wheat is about 1,200 ppm,[16] nearly four times the current levels! In many plant studies that are designed to test plant growth under "optimum" conditions, carbon dioxide levels are controlled at about 1,000 ppm. It is no wonder that hundreds of commercial nurseries routinely add carbon dioxide to the atmosphere inside greenhouse operations to enhance the growth and yield of plants.

Scientific evidence also shows that plants utilize water more efficiently when exposed to elevated carbon dioxide levels, according to Professor Cure.[17] This effect is illustrated in the "moisture

Table 6.2. Summary of Doubling the Carbon Dioxide Levels on Several Major Farm Crops[a]

Crop & Species	World rank	% Biomass accumulation[b]	% Moisture transpiration[b]	% Yield[b]
Wheat	1	+31±16	−17±17	+35±14
Triticum aestivum L.		(10)	(2)	(8)
Barley	4	+30±17	−19±6	+70±[c]
Hordeum vulgare L.		(6)	(3)	(1)
Rice	2	+27±7	−16±9	+15±3
Oryza sativa L.		(11)	(3)	(3)
Corn	3	+9±5	−26±6	+29±64
Zea mays L.		(13)	(6)	(1)
Sorghum	6	+9±29	−27±16	—
Sorghum bicolor (L.) Moench.		(3)	(2)	
Soybean	7	+39±5	−23±5	+29±8
Glycine max (L.) Merr.		(20)	(8)	(12)
Cotton	9	+84±126	−18±17	+209±[c]
Gossypium hirsutum L.		(3)	(3)	(1)
Potato	12	−15±4	−51±24	+51±111
Solanum tuberosum L.		(1)	(3)	(3)

[a]Data from Jennifer D. Cure, "Carbon Dioxide Doubling Responses: A Crop Survey," *Direct Effects of Increasing Carbon Dioxide on Vegetation*, Boyd R. Strain and Jennifer D. Cure, eds. (Washington, D.C.: U.S. Department of Energy, DOE/ER-0238, 1985), pp. 103–105. The values in parentheses are the number of studies used in calculating the values as stated.

[b]Percentage change at 680 ppm CO_2 compared with controls (300–350 ppm) ±95% confidence limits, as estimated by regression analysis. Biomass refers to the weight of the plants; moisture transpiration to the amount of water transmitted to the atmosphere, and yield to the amount of the crop produced per acre.

[c]Insufficient data to calculate a confidence limit.

transpiration" column of Table 6.2. The decrease in moisture transpiration indicates that less water vapor is emitted from the plant's leaves. This means that as the concentration of carbon dioxide goes up, crops need less water for the same yield level. The implication is that crops could possibly be grown in arid areas that would not have supported them in the past, due to insufficient water supplies. Deserts could become producing farmlands! Areas that have growing populations and farmers competing for water supplies could sustain both for a longer period of time. Maybe agriculture will be able to keep up with our out-of-control population growth for a few extra years!

OTHER GREENHOUSE GASES

And what about the other greenhouse gases? Methane, nitrous oxide, chlorofluorocarbons, carbon monoxide, and ozone are all included in discussions of greenhouse gases. During the decade of the 1980s they increased as well, but there has been a leveling off of most of these gases and even a downturn in atmospheric concentration for many of them since the 1990s.

These gases exist in the atmosphere at much lower levels than carbon dioxide. Whereas the current level of carbon dioxide is about 0.035% by volume, which is 350 ppm by volume, the level of methane gas (CH_4) is only 0.00017% (1.7 ppm). The composition of the (dry) atmosphere, including the relative amounts of some of the other chemical compounds that contribute to the greenhouse effect, is shown in Table 6.3. If carbon dioxide is a minor component of the atmosphere, then all other greenhouse gases must truly be considered as trace components because their concentration in the atmosphere is extremely low.

These gases vary in their level of efficiency in the greenhouse effect because they interact to differing degrees with the infrared radiation being emitted by the earth. The greater this interaction, the more effective the gas is as a greenhouse gas. Table 6.4 gives the relative effect of various greenhouse gases with respect to carbon dioxide. This is known as the global warming potential (GWP). The GWP has been used as a measure of the importance of various gases in relation to their effectiveness as greenhouse gases. For example, the GWP has become an important criterion in finding replacements for CFCs. As you can see from Table 6.4, carbon dioxide is one of the *least* efficient greenhouse gases.

When considered in the global warming computer models, each of the trace components' concentration values is multiplied by the GWP factor in Table 6.4 to make its effect equal to that of carbon dioxide. In this way all of the greenhouse gases are summed

Table 6.3. Composition of the Atmosphere

Gas	Chemical formula	Green-house gas	Abundance percent	ppmv[a]	ppbv[b]	Ref.
Nitrogen	N_2	no	78.08	—	—	1
Oxygen	O_2	no	20.95	—	—	1
Argon	Ar	no	0.93	930	—	1
Water	H_2O	yes	1–3	—	—	2
Carbon Dioxide	CO_2	yes	0.035	350	—	3
Ozone	O_3	yes	—	0.03–10	—	1
Methane	CH_4	yes	—	1.7	1700	3
Neon	Ne	no	—	18	18,000	1
Helium	He	no	—	5	5000	1
Krypton	Kr	no	—	1	1000	1
Xenon	Xe	no	—	0.08	80	1
Hydrogen	H_2	no	—	0.5	500	1
Nitrous Oxide	N_2O	yes	—	0.3	300	3
Carbon Monoxide	CO	yes	—	0.070	70	4
Sulfur Dioxide	SO_2	?[c]	—	0.001	1	1
Nitrogen Dioxide	NO_2	yes	—	0.001	1	1
Ammonia	NH_3	yes	—	0.004	4	1
Hydrogen Sulfide	H_2S	yes	—	—	0.5	1

[a]ppmv means part per million by volume or one molecule of the gas per million (10^6) molecules of total atmosphere.

[b]ppbv means parts per billion by volume means one molecule per billion (10^9) molecules of total atmosphere.

[c]The SO_2 molecules are precursors for cloud formation and sulfate particulates, both of which cause cooling (see Chapter 9).

[1]Melvin A. Benarde, *Global Warning . . . Global Warming* (New York: John Wiley & Sons, 1992) p. 46.

[2]Stanley E. Manahan, *Environmental Chemistry*, 6th. ed. (Roca Raton, LA: Lewis Publishers, 1994) p.31. Because of the large regional variation in the atmosphere of water, the other constituents are generally calculated on a "dry air" basis.

[3]Thomas A. Boden, Robert J. Sepanski, and Frederick W. Stoss, eds. *Trends '91: A Compendium of Data on Global Change* (Oak Ridge: U.S. Department of Energy, ORNL/CDIAC-46) p. 15 for CO_2 and p. 235 for CH_4.

[4]M. A. K. Khaill & R. A. Rasmussen, "Global decrease in atmospheric carbon monoxide concentration," *Nature*, 370 (1994) pp. 639–641.

Table 6.4. Relative Global Warming Efficiency Compared to Carbon Dioxide (Global Warming Potential, GWP)[a]

Gas	Chemical formula	Lifetime[b] (years)	Direct effect[c] (years)	Indirect effect[d]
Carbon Dioxide	CO_2	about 120	1	none
Methane	CH_4	10.5	35	positive
Nitrous Oxide	N_2O	132	260	uncertain
Chlorofluorocarbons	CFCs	55→500	4500–6100	negative
Carbon monoxide	CO	months	0	positive
Ozone	O_3	short	0	unknown
Other nitrogen oxides	NO_x	days	0	unknown

[a]Adapted from R. T. Watson, L. G. Meira Filho, E. Sanhueza, and A. Janetos, "Greenhouse Gases: Sources and Sinks," *Climate Change 1992*, J. T. Houghton, B. A. Callander, and S. K. Varney, eds. (Cambridge: Cambridge University Press, 1992), p. 56.

[b]This refers to the life of the species in the atmosphere. These lifetimes are not well known; the ones presented here are those used by the IPCC to calculate the global warming results presented here.

[c]These calculated values for the direct global warming effect vary with time. The ones presented here are for 20 years in the future.

[d]A *positive* indirect effect means that the effect will be greater than stated due to other processes than the greenhouse effect. A *negative* indirect effect means that the effect will be less than stated due to other processes than the greenhouse effect. *None* means that no known indirect effects will take place.

into an equivalent carbon dioxide concentration, which simplifies the computation of the total effect of the greenhouse gases in the computer models. Although this is the common practice of the global modeling community, it does not account accurately for the variation in the chemical behavior of each gas in the atmosphere. This procedure accounts for the direct effects, but not for any indirect effects, such as chemical reactions with other gases in the atmosphere. The indirect interactions of each of these gases are quite different, and these differences are not being accounted for in the computer models.

Methane

After carbon dioxide, methane is the next most commonly maligned of the greenhouse gases. It has tripled from its historical

Table 6.5. Sources for Methane in the Atmosphere[a]

Natural sources	Range[b]	IPCC value[c]
Wetlands	100–200	115
Termites	10–50	20
Ocean	5–20	10
Freshwater	1–25	5
Methane hydrates	0–5	5
Subtotals	116–300	155
Anthropogenic sources		
Coal, natural gas, & petroleum industry	70–120	100
Rice paddies	20–150	60
Enteric fermentation[d]	65–100	80
Animal wastes	20–30	25
Domestic sewage treatment	?	25
Landfills	20–70	30
Biomass burning	20–80	40
Subtotals	215–550	360

[a]Adapted from R. T. Watson, L. G. Meira Filho, E. Sanhueza, and A. Janetos, "Greenhouse Gases: Sources and Sinks," *Climate Change 1992*, J. T. Houghton, B. A. Callander, and S. K. Varney, eds. (Cambridge: Cambridge University Press, 1992), p. 35.

[b]As this information is only poorly known, the range of data is provided here. The units are in teragrams (10^{12} g). One teragram = 1 megaton (10^6 t). One megaton is 1,000,000 tons.

[c]This is the value deemed most likely by the IPCC.

[d]This constitutes gaseous emissions from animals. The IPCC considers all animals as being caused by activities of mankind.

baseline, or background, value of 0.3–0.6 ppm some 200 to 300 years ago, as determined by measurements of trapped ice core gases. The sources and sinks for methane are even less well known than those of carbon dioxide. About half of the sources are considered to be "natural," as indicated in Table 6.5. One of the major anthropogenic sources is "enteric fermentation"—this is flatulence, the gas generated by the digestive systems of virtually all mammals, including humans. The total wild animal population of the earth must surely account for a large percentage of this category and should be included under natural sources, but it is not. The numbers in Table 6.5 are so poorly known that large ranges of numbers must be used.

Many sources of methane are associated with the human population of the earth—domestic animals, rice paddies, biomass burning, landfills, coal mining, and gas drilling. The exponential growth of the world's population will cause an increase in the amount of methane in the atmosphere.

It is interesting to note that the measured concentration of methane in the atmosphere has recently taken a downward turn, but scientists are not sure of the reasons. Richard A. Kerr, reporter for *Science* magazine, lists the possibilities:

- Fewer leaks in Russian pipelines because of repairs prompted by a major gas explosion in 1989.
- The eruption of Mt. Pinatubo and its subsequent cooling effect, which in turn have decreased the methane emissions from wetlands.
- Unknown causes.

In his report, Kerr quotes Ralph Cicerone, well-known atmospheric chemist at the University of California, Irvine, as stating that atmospheric methane represents "a system with a number of unknowns and equations, and I'm rather sure that there are more unknowns than equations."[18] Once again, the scientists who are the core of the climate research community admit that the uncertainties outweigh the certainties when dealing with our complex atmosphere.

Chlorofluorocarbons (CFCs)

Most CFCs are man-made chemicals that were first developed in the 1930s, and they have had a major impact on society and our quality of life. Chlorofluorocarbons are used as refrigeration gases in air conditioning, as solvents and foaming agents in industrial processes, and as fire suppression agents in firefighting, as well as in many other applications. Although CFCs are usually included in discussions of global warming, they are probably no longer a real consideration because they have been banned by the Montreal Protocol treaty. This ban was not imposed because of their global warming potential but because of their perceived role in the depletion of ozone in the stratosphere. Later, their role as greenhouse gases was recognized.

The production of CFCs has been stopped or will be stopped in the very near future all over the world, with very small production allowed, and only where there is a "critical" use, typically in one of the underdeveloped nations. There are several uses for CFCs that are considered critical. For example, there have been no replacements found for the life-saving use of certain halon firefighting agents in areas in which human occupation is essential during the firefighting process. These include aircraft, military tanks, and any facility from which people cannot escape the fire area without further endangering their lives, such as oil rig platforms in the northern oceans, facilities on the northern slopes of Alaska, or in Antarctica. The military has stockpiled these firefighting agents and plans to use recycling procedures to prolong their useful life.

The Montreal Protocol was driven by results of computer models that are in many ways similar to those used by the global warming scientists. A major difference is that the ozone models include predictions of chemical reactions in the stratosphere in conjunction with circulation models that predict the transport of CFCs and other gases up to the level of the stratosphere. The computer climate models have generally ignored the effects of chemical reactions in the atmosphere.

An interesting twist with respect to overall climatic effects caused by CFCs was recently introduced by V. Ramaswarmy and M. D. Schwarzkopf of Princeton University and K. P. Shine of the University of Reading in England.[19] They concluded that the ozone depletion caused by CFCs has resulted in a global cooling effect, which in large part offsets any global warming effect of these chemicals. In spite of this lack of understanding as to the warming or cooling effects of CFCs, global warming potential (GWP) is a major criterion for the selection of CFC replacement chemicals.

Nitrous Oxide (N_2O)

Table 6.6 provides the current information on the sources of nitrous oxide. The major natural sources of nitrous oxide are the oceans and forests. The largest anthropogenic source of nitrous oxide appears to be soils on cultivated farmlands; however, if the true values for anthropogenic sources are at the low end of the

Table 6.6. Sources for Nitrous Oxide in the Atmosphere[a]

Natural sources	Range[b]
Oceans	1.4–2.6
Tropical soils	?
Wet forests	2.2–3.7
Dry savannas	0.5–2.0
Temperate soils	?
Forests	0.5–2.0
Grasslands	?
Anthropogenic sources	
Cultivated soils	0.3–3.0
Biomass burning	0.03–1.0
Combustion	0.1–0.3
Mobil sources	0.2–0.6
Adipic acid production	0.4–0.6
Nitric acid production	0.1–0.3
Totals given by IPCC	5.2–16.1[c]

[a]Adapted from R. T. Watson, L. G. Meira Filho, E. Sanhueza, and A. Janetos, "Greenhouse Gases: Sources and Sinks," *Climate Change 1992*, J. T. Houghton, B. A. Callander, and S. K. Varney, eds. (Cambridge: Cambridge University Press, 1992), pp. 37–38.

[b]As this information is only poorly known, the range of data is provided here. The units are in teragrams (10^{12} g). One teragram = 1 megaton (10^6 t). One megaton is 1,000,000 tons.

[c]These totals were deduced from the magnitude of the sinks and the rate of accumulation in the atmosphere.

spread of numbers given in Table 6.6, these anthropogenic sources are not really of major importance. These numbers are very uncertain and were modified rather substantially by the IPCC between their 1990 and 1992 reports.

Nitrous oxide is another trace greenhouse gas. You may have heard of it as "laughing gas," an anesthetic used by dentists. As seen from Table 6.3, nitrous oxide is present in the atmosphere at only about one-sixth the level of methane; however, it has about seven times the direct GWP (Table 6.4).

The atmospheric chemistry of nitrous oxide and its role in indirect climate change is uncertain. This uncertainty was highlighted in a recent study by a large group of scientists led by Dr. P. O. Wennberg, chemist at Harvard University. This research team included scientists from the National Oceanic and Atmospheric Administration Aeronomy Laboratory, the Jet Propulsion Laboratory, the University of California at Irvine, and NASA's Ames Research Center. The study suggests that nitrous oxide does not dominate the chemical destruction of ozone as had been thought for the past two decades.[20] The findings of these researchers, based on measurement of species in the stratosphere, not on computer models, will create a major change in the computer models of stratospheric chemistry. The warming and/or cooling in the stratosphere has a direct effect on the warming and cooling in the troposphere. The chemistry of the stratosphere will ultimately have to be included in global climate models if they are to be complete.

Ozone

Ozone is a very reactive form of oxygen that attacks most chemicals in the atmosphere and has a very short lifetime. It is considered to be an air pollutant because it irritates the lungs, and it is thus controlled by the Clean Air Act. Ozone is a greenhouse gas, but it is present in such low quantities and has such a short lifetime in the atmosphere that its main effects are indirect and not well understood. Its complex chemistry is incompletely known both in the troposphere and the stratosphere. As a consequence, its overall effect in the atmosphere is very uncertain. Nevertheless, ozone is always included in discussions of greenhouse gases. People often confuse the ozone layer in the stratosphere, where it absorbs high-energy (harmful) radiation from the sun, and the ozone in the troposphere, where it is generally considered an air pollutant.

Like carbon dioxide, methane, and the other greenhouse gases, there is incomplete knowledge with respect to the sources and sinks of ozone, and computer models indicate that there is more ozone in the air than is actually observed. Recently, a group of scientists led by Professor R. L. Miller of the Chemistry Department at Cornell University found new chemical reactions involving ozone that led to a closer agreement between prediction and reality.[21] The fact that scientists are finding new chemical reactions in the atmosphere involving ozone, nitrous oxide, and other greenhouse gases is of utmost importance to computer climate models. These reactions have not yet been included in the models, and incomplete inputs into the models mean incomplete outputs.

The IPCC scientists thought they had enough understanding of ozone chemistry to warrant inclusion of the ozone data reported in their 1990 report. However, they removed this data from the 1992 report,[22] admitting that there are too many complexities for a real understanding of ozone and other trace-level gases such as carbon monoxide and other oxides of nitrogen with which ozone reacts.

An interesting side note is the fact that carbon monoxide, one of the short-lived precursors of ozone, which was thought to be increasing due to man's activities, is currently decreasing in the atmosphere, according to measurements made by M. A. K. Khalil and R. A. Rasmussen of the Global Change Research Center of the Oregon Graduate Institute.[23] It had been thought that all the trace gases in the atmosphere were increasing. However, the concentrations of many of these gases have been slowing down and even decreasing. The slowing trends are being observed in the concentrations of carbon dioxide, methane, and some of the oxides of nitrogen, in addition to carbon monoxide. The reasons for these slowing trends are a mystery. Some scientists blame the volcanic activity of Mt. Pinatubo. Others scientists blame the weather. Regardless of the causes, the slowing trends were not predicted by the computer climate models.

WHAT HAVE WE LEARNED?

We can conclude several things from this chapter. There is no question that the greenhouse gas phenomenon exists! If it did not, the earth would be a frozen wasteland. Carbon dioxide, the major man-made greenhouse gas (excluding water vapor), is increasing in the atmosphere. Other greenhouse gases, such as methane, also have been increasing; however, there has recently been a leveling in their increase and in some cases even a downturn in concentration.

If we look at the historical levels of atmospheric carbon dioxide and methane, we find that they have been higher when the earth was warmer and lower when the earth was colder. It is possible that the variation in greenhouse gases was in large measure caused by the variation in the temperature of the earth, and not the other way around.

Finally, we must answer the question: Is an increase in the carbon dioxide concentration in the earth's atmosphere a bad thing? Carbon dioxide clearly helps the growth of plant life. Most agricultural scientists spend their lives developing ways to improve crop yields. Maybe Arrhenius was right—maybe a warmer world is a better world.

REFERENCES

1. M. J. Fisher, I. M. Rao, M. A. Ayarza, C. E. Lascano, J. I. Sanz, R. J. Thomas, and R. R. Vera, "Carbon Storage by Introduced Deep-rooted Grasses in the South American Savannas," *Nature* 371 (1994):236–238.
2. Daniel C. Nepstad, Claudio R. de Carvalho, Eric A. Davidson, Peter H. Jipp, Paul A. Lefebvre, Gustavo H. Negreiros, Elson D. da Silva, Thomas A. Stone, Susan E. Trumbore, and Simone Vieira, "The Role of Deep Roots in the Hydrological and Carbon Cycles of Amazonian Forests and Pastures," *Nature* 372 (1994):666–669.

3. Anthony F. Michaels, Nicholas R. Bates, Ken O. Buesseler, Craig A. Carlson, and Anthony H. Knap, "Carbon-Cycle Imbalances in the Sargasso Sea," *Nature* 372 (1994):537–540.
4. John R. Trabalka, ed., *Atmospheric Carbon Dioxide and the Global Carbon Cycle* (Oak Ridge, TN: U.S. Department of Energy, DOE/ER-0239, 1985), pp. xv–xxiii.
5. U. Siegenthaler and J. L. Sarmiento, "Atmospheric Carbon Dioxide and the Ocean," *Nature* 365 (1993):119–127.
6. Sherwood B. Idso, "Carbon Dioxide and Climate in the Vostok Ice Core," *Atmos. Environ.* 22 (1989):2341–2342.
7. William James Burroughs, *Weather Cycles: Real or Imaginary?* (Cambridge: Cambridge University Press, 1992), pp. 142–143.
8. D. C. Keeling and T. P. Whorf, "Atmospheric CO_2 Records from Sites in the SIO Air Sampling Network," in *Trends '93: A Compendium of Data on Global Change*, ORNL/CDIAC-65, ed. T. A. Boden, D. P. Kaiser, R. J. Sepanski, and F. W. Stoss (Oak Ridge, TN: U.S. Department of Energy, 1994), pp. 16–26.
9. J. M. Barnola, D. Raynaud, C. Lorius, and Y. S. Korotkevich, "Historical CO_2 Record from the Vostok Ice Core," in *Trends '93: A Compendium of Data on Global Change*, ORNL/CDIAC-65, ed. T. A. Boden, D. P. Kaiser, R. J. Sepanski, and F. W. Stoss (Oak Ridge, TN: U.S. Department of Energy, 1994), pp. 7–10.
10. A. Neftel, H. Friedli, E. Moor, H. Lotscher, H. Oeschger, U. Siegenthaler, and B. Stauffer, "Historical CO_2 Record from the Siple Station Ice Core," in *Trends '93: A Compendium of Data on Global Change*, ORNL/CDIAC-65, ed. T. A. Boden, D. P. Kaiser, R. J. Sepanski, and F. W. Stoss (Oak Ridge, TN: U.S. Department of Energy, 1994), pp. 11–14.
11. R. H. Gammon, E. T. Sundquist, and P. J. Fraser, "History of Carbon Dioxide in the Atmosphere," in *Atmospheric Carbon Dioxide and the Global Carbon Cycle*, DOE/ER-0239, ed. John R. Trabalka (Oak Ridge, TN: Department of Energy, 1985), pp. 27–28.
12. Harold W. Bernard, Jr., *Global Warming Unchecked: Signs to Watch For* (Bloomington, IN: Indiana University Press, 1993), p. 5.
13. Sherwood B. Idso, *Carbon Dioxide and Global Change: Earth in Transition* (Tempe, AZ: IBR Press), pp. 67–89.
14. B. A. Kimball, *Carbon Dioxide and Agricultural Yield: An Assemblage and Analysis of 770 Prior Observations* (Phoenix, AZ: U. S. Water Conservation Laboratory, 1983).
15. Jennifer D. Cure, "Carbon Dioxide Doubling Responses: A Crop Survey," in *Direct Effects of Increasing Carbon Dioxide on Vegetation*, DOE/ER-0238, ed. Boyd R. Strain and Jennifer D. Cure (Washington, DC: U.S. Department of Energy, 1985), pp. 101–116.
16. B. G. Bugbee and F. B. Salisbury, "Exploring the Limits of Crop Productivity. I. Photosynthetic Efficiency of Wheat in High Irradiance Environment," *Plant Physiol.* 88 (1988):869–878.
17. Cure, 112.

18. Richard A. Kerr, "Methane Increase Put on Pause," *Science* 263 (1994):751.
19. V. Ramaswarmy, M. D. Schwarzkopf, and K. P. Shine, "Radiative Forcing of Climate from Halocarbon-Induced Global Stratospheric Ozone Loss," *Nature* 355 (1992):810–812.
20. P. O. Wennberg, R. C. Cohen, R. M. Stimpfle, J. P. Koplow, J. G. Anderson, R. J. Salawitch, D. W. Fahey, E. L. Woodbridge, E. R. Keim, R. S. Gao, C. R. Webster, R. D. May, D. W. Toohey, L. M. Avallone, M. H. Proffitt, M. Loewenstein, J. R. Podolske, K. R. Chan, S. C. Wofsy, "Removal of Stratospheric O_3 by Radicals: In Situ Measurements of OH, HO_2, NO_2, ClO, and BrO," *Science* 266 (1994):398–404.
21. R. L. Miller, A. G. Suits, P. L. Houston, R. Toumi, J. A. Mack, A. M. Wodtke, "The Ozone Deficit Problem: $O_2(X, v \geq 26) + O(^3P)$ from 226-nm Ozone Photodissociation," *Science* 265 (1994):1831–1838.
22. R. T. Watson, L. G. Meira Filho, E. Sanhueza, and A. Janetos, "Greenhouse Gases: Sources and Sinks," in *Climate Change 1992*, ed. J. T. Houghton, B. A. Callander, and S. K. Varney (Cambridge: Cambridge University Press, 1992), p. 40.
23. M. A. K. Khalil and R. A. Rasmussen, "Global Decrease in Atmospheric Carbon Monoxide Concentration," *Nature* 370 (1994):639–641.

SEVEN

The Sun

And the Moon in haste eclipsed her,
and the Sun in anger swore
He would curl his wick within him
and give light to you no more.
—*Aristophanes, The Clouds*

Through the ages people have recognized the importance of the sun. In culture after culture sun deities have been worshipped: The Greeks had their Helios; the Aztecs, their Temple of the Sun; and the American Plains Indians, their sun dances. The sun has intrigued and mystified astronomers throughout history, from Ptolemy in the second century A.D. to today's astrophysicists and solar spectroscopists.

With the invention of the telescope early in the seventeenth century, knowledge about the universe grew, and people eventually understood that the sun, while being the center of our solar system, is only one of many billions of stars in the universe. Stars are great boiling infernos fed by nuclear reactions of hydrogen and helium gas; they come in all sizes and colors and emit various types of radiation and energy. The discussion below will concentrate on the star that warms the earth, our sun.

The sun sends out a continuous energy to the solar system, which includes the planets, their moons, asteroids, and a wide assortment of comets. This continuous energy is often thought to be a constant, but in reality the sun's behavior is chaotic, and its energy output is ever changing. It is only since the advent of the space age that scientists have had the ability to obtain accurate data on this chaotic behavior. The earth's atmosphere, which provides a shield from the sun's deadly high-energy radiation, also acts as an observational barrier that kept valuable information inaccessible until the space program made it possible to place solar probes beyond the atmosphere. Now sensitive instruments are placed in space with the purpose of gathering new information about the sun.

SUNSPOTS

One of the features of the sun that is indicative of its chaotic behavior is the appearance of sunspots, which can be seen from earth. Galileo, one of the first scientists to use a telescope, studied the sun in the early 1600s. He was fascinated by the dark spots that moved across its surface and concluded that the sun was a rotating sphere. His experimental study provided an interpretation of the sun that holds up today. He noted that shapes, sizes, and lifetimes of the sunspots were quite variable. Galileo used the earth's clouds as an analogy in his description of the sunspots, but he did not have sufficient observational tools to know if the sun had a gaseous atmosphere; he had no real idea as to the physical nature of the sunspots.

During the next two centuries, sunspots, the only visible evidence of the sun's internal activity, were observed carefully by astronomers. One of these astronomers was Sir John Herschel, of Cambridge, England. Herschel's father, Sir William Herschel, was an astronomer and telescope builder who discovered hundreds of stars. William Herschel, with the help of his sister, Caroline, catalogued all of the visible stars in the Northern Hemisphere along with 2,500 nebulae and 848 double stars. Following in his famous father's footsteps, John Herschel pursued astronomy; he and his family spent four years at the Cape of Good Hope and catalogued stars in the Southern Hemisphere that could not be observed from

England. He provided a major catalogue of the stars in the Southern Hemisphere. Ultimately, he became interested in solar astronomy and studied sunspots in the early nineteenth century. In his book *Outlines of Astronomy*, which was published in 1849, he described sunspots in this way:

> When viewed through powerful telescopes, provided with coloured glasses, to take off the heat, which would otherwise injure our eyes, the sun is observed to have frequently large and perfectly black spots upon it, surrounded with a kind of border, less completely dark, called penumbra. . . [1]

He described the movements and variations of the sunspots, which are not permanent. He noted that sunspots enlarge and contract, change their shapes, and then disappear. When they disappear, the central dark spot always contracts into a point and vanishes. Occasionally they break up, or divide into two or more, and in those cases they offer every evidence of the extreme mobility that is symptomatic of the fluid or gaseous state, as opposed to a solid. There are often signs of violent agitation, which seems compatible with only the gaseous, or atmospheric, state of matter. Herschel commented on the magnitude and speed of the sunspots and their motion:

> The scale on which their movements take place is immense. A single second of angular measure, as seen from the earth, corresponds on the sun's disc to 461 miles; and a circle of this diameter (containing therefore nearly 167,000 square miles) is the least space which can be distinctly discerned on the sun as a visible area. Spots have been observed, however, whose linear diameter has been upwards of 45,000 miles; and even if some records are to be trusted, of very much greater extent. That such a spot should close up in six weeks' time (for they seldom last much longer), its borders must approach at the rate of more than a 1000 miles a day.[2]

A sunspot 45,000 miles in diameter could swallow the earth, which is only about 8,000 miles in diameter! The sun is so large that a sunspot of 45,000 miles still looks like a little black spot. When put in the earth-sized scale of things that humans normally consider, sunspots are awesome in both size and dynamics of formation and disappearance.

Herschel's colorful description gives a feeling for the dramatic visual action associated with sunspots, as well as the continuous boiling activity throughout the surface of the sun. Herschel thought that the sun had a luminous, gaseous outer layer and a solid inner surface; hence, he spoke of the "ground" beneath the sunspots. (We know today that the sun has no solid surface.) Herschel rightly concluded that the sun's energy output, and therefore the earth's temperature, was affected by sunspot activity.[3] Much of what Herschel had to say about the sun was pure speculation. Little was known about the true source of the sun's energy or the solar chemistry that creates that energy.

Although Herschel was not aware of it at the time, an amateur astronomer by the name of Heinrich Schwabe had been making daily observations of sunspots. According to solar scientist Kenneth Phillips, in his book *Guide to the Sun*, Schwabe, a prosperous pharmacist from Germany, got so involved with his sunspot study that he ultimately sold his pharmacy and became a full-time astronomer.[4] After more than twenty years of detailed observations, he concluded that there was a sunspot cycle with an average duration of ten years. This was the first discovery of the solar cycle. Schwabe published his studies in 1843 in a German astronomical journal, where they attracted little attention until J. Rudolf Wolf, of Bern University in Switzerland and later the director of the Zurich Observatory, gathered all the sunspot information from Galileo's time up through Schwabe's study. Wolf concluded that there was indeed a sunspot cycle, but that its average duration was eleven years rather than ten. His data and conclusions, published in 1877, provided the scientific community with the seeds of understanding of the true nature of the sun's activity and variability. Wolf was famous for his solar studies, and the methods and procedures he used, slightly improved, are still in use by astronomers today. Wolf also discovered that disturbances in the earth's magnetic field were associated with sunspot activity, thus linking this visible activity with much greater magnetic activity felt on the earth and throughout the solar system.

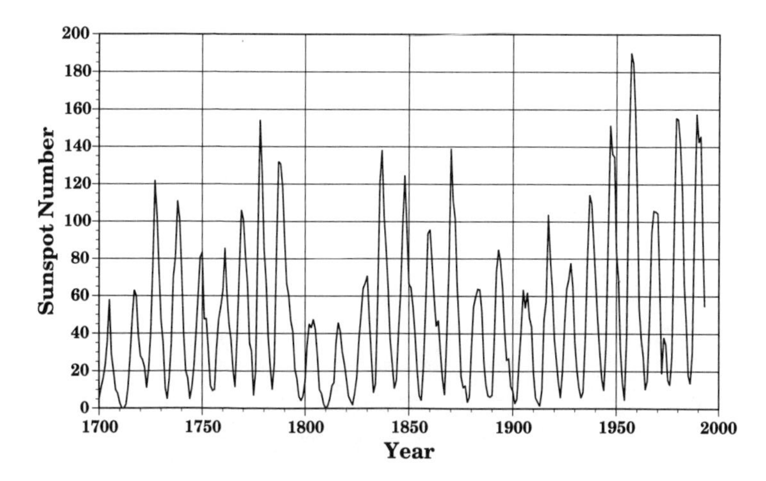

FIGURE 7.1. Sunspot numbers from 1700 to 1993. The data for this figure were taken from John A. McKinnon, *Sunspot Numbers: 1610–1985* (Boulder, CO; World Data Center A for Solar-Terrestrial Physics, 1987), plus updates.

Like many natural phenomena, the sunspot cycle, illustrated in Figure 7.1, does not produce a pure cycle and is quite variable. The periods range in length from about seven years (between the maxima, or peaks, of 1830 and 1837) to sixteen years (between the maxima of 1788 and 1804). In the past century, during which time the observations have been more consistent, the cycles have ranged from nine to twelve years. The intensity of sunspots varies from maxima to maxima as well. Between 1650 and 1700 there was little or no sunspot activity. This period, referred to as the Maunder minimum, corresponds with the Little Ice Age.

In the late nineteenth century, Charles A. Young, professor of physics and astronomy at Dartmouth College, was engaged in major research on the nature of the sun. He suggested the likelihood of a longer cycle in addition to the eleven-year cycle. Young stated in his book *The Sun*: "It is quite likely that a fluctuation of much longer period, not far from sixty years, is to some extent, responsible for the effect [of variation of the cycle maxima] by its

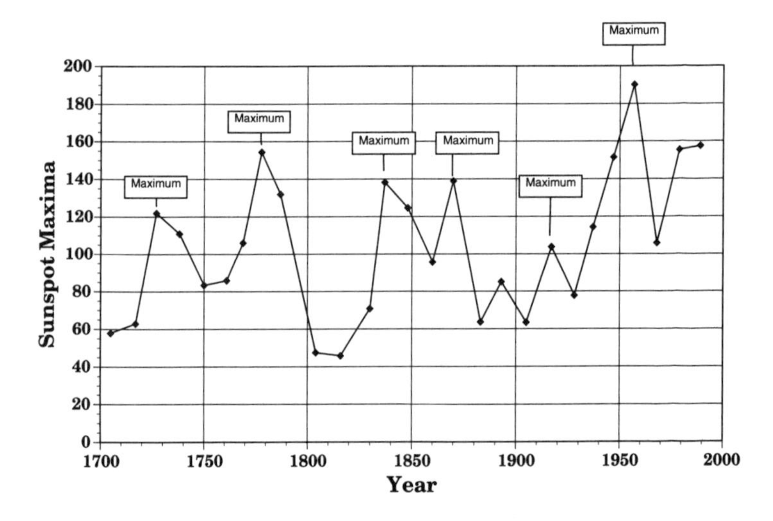

FIGURE 7.2. Sunspot maxima from 1700 to 1993. The data for this figure were taken from John A. McKinnon, *Sunspot Numbers: 1610–1985* (Boulder, CO; World Data Center A for Solar-Terrestrial Physics, 1987), plus updates.

superposition upon the principal (eleven-year) oscillation."[5] This longer cycle is easily observed if the maxima of each sunspot cycle is plotted versus time, as shown in Figure 7.2. The average period indicated by the maxima in Figure 7.2 would make this cycle about forty to fifty years instead of the sixty predicted by Young. He did not have the advantage of sunspot information from the past century.

Long-term cycles or not, Herschel's belief was confirmed; the sunspots were indeed related to the intensity of radiation and hence the heat derived from the sun. By the end of the nineteenth century, the eleven-year solar cycle had become an accepted fact of solar physics.

THE SOLAR CONSTANT

A question puzzled the scientists of the 1800s: "What is the source of the sun's energy, how does it keep going, and why doesn't its energy fade away?" Herschel asked the question in his *Outlines of Astronomy* as follows:

> The great mystery, however, is to conceive how so enormous a conflagration (if such it be) can be kept up. Every discovery in chemical science here leaves us completely at a loss, or rather, seems to remove farther the prospect of probable explanation. If conjecture might be hazarded, we should look rather to the known possibility of an indefinite generation of heat by friction, or to its excitement by the electric discharge, than to any actual combustion of ponderable fuel, whether solid or gaseous, for the origin of the solar radiation.[6]

It is interesting that Herschel questioned and rejected the concept of a consumable fuel feeding the sun's furnace. He had a reasonable idea of the enormity of the sun and had calculations as to its massive size. Perhaps the concept of a sun that could ultimately burn up was more than he wanted to understand. His idea about heat generated by friction seems ill thought out. If the sun generates its heat by means of friction as it hurtles through space, why wouldn't the earth also be subject to this same friction? Finally, since the metal of an electric welding arc is consumed in the process of welding, it would be logical to think that an electric discharge would also result in the consumption of some type of fuel. However, his idea of an electrical-type discharge is very interesting because electric arcs are one of the few methods by which scientists can create sunlike temperatures on earth.

This question of ultimate sun burnout prompted researchers to attempt to measure the energy received on the earth from the sun. Professor Hufbauer, in his book *Exploring the Sun*, relates that Claude Pouillet, a nineteenth-century French physicist, defined and measured the "solar constant." According to his definition, the solar constant is the amount of energy from the sun impinging on one square centimeter (cm^2) of the earth's outer atmosphere.[7] A square centimeter is an area about the size of the end of an eraser on a pencil. Because his measurements were made from the earth's surface rather than above the atmosphere, he realized that the value was not truly accurate, but his measurement of about 1.8 calories per square centimeter per minute ($cal/cm^2/min$) was surprisingly good. The currently accepted value for the solar constant is very close to 2 cal/cm^2/min. Put into more understandable units, the solar constant provides about 1,360 watts of energy for every square meter,[8] or the energy produced by some fourteen 100-watt light bulbs on a square surface of 1 meter on each side. Of course not all of that energy reaches the earth; it must pass through the atmosphere first.

It is unfortunate that Pouillet used the term "constant" in his definition; neither he nor most scientists of his day actually thought that the sun's radiation output was truly a constant. Nevertheless, the implication of a constant solar output was there, and the sun's energy has been considered a constant in most of the current global warming computer models. This is usually done to save expensive computational time and because of the belief among many climate scientists that solar variation is too small to be concerned about. If small variable effects are omitted from computer climate models, the complicated mathematical interactions in the models can create a domino effect causing major errors in the final results. All variables should be included for accurate results in a computer model.

UNDERSTANDING THE SUN'S CHEMISTRY

The field of solar physics blossomed during the 1800s, and the solar spectrum, the intensity of solar radiation as a function of its wavelength, was a major area of scientific study. A fundamental understanding of the solar spectrum would be necessary before the nature of the solar constant could be determined with accuracy. Charles Young described the progression of discovery in *The Sun*.[9] In 1814 an optician in Munich, Joseph Fraunhofer, observed the sun through a spectroscope and discovered hundreds of dark lines in the visible spectrum. Although an English physicist, William Wollaston, had first observed these dark shadings in 1802, it was Fraunhofer who "mapped" the lines and for whom they were named, and he created a table that showed the relative position of the dark lines.

This discovery was a great puzzle to solar scientists until physicist Gustav Kirchhoff, working with his colleague Robert Bunsen at Heidelberg University in 1859, developed the spectroscopic tools and procedures that would enable them to observe the sun's radiation with high precision and help them to solve the mystery. (One of these tools was the Bunsen burner, in use today in science laboratories throughout the world.) Bunsen and Kirchhoff determined that the Fraunhofer lines were coincidental with atomic emission lines from specific elements. Atomic emission lines are the "fingerprints" of the elements; they are composed of the radiation emitted by atoms of an element when they are excited in the flame of a Bunsen burner (or the sun). Kirchhoff and Bunsen re-created the phenomenon of Fraunhofer lines in their laboratory and determined that they were caused by chemical elements present in the sun's atmosphere. The chemical elements absorbed their fingerprint radiation (the dark lines) from the sun's continuum radiation. It was now possible to identify the chemical composition of the sun, and according to Young, Kirchhoff set out to do so. Kirchhoff identified the presence of several elements in the solar spectrum, and subsequent studies by many scientists have confirmed the presence of most of the elements that make up the modern-day periodic table of the elements.

The question of the sun's longevity and energy output could not yet be answered, however, because no one knew the source of its energy. During the nineteenth century there were two speculative theories. According to Professor Hufbauer in *Exploring the Sun,* one group thought that gravitational forces coupled with a shrinking sun provided the energy, and the other group was sure that objects such as meteors hitting the sun were the fuel for the furnace.[10] Two missing puzzle pieces had yet to be discovered: the quantities of the elements present in the sun and an understanding of the solar chemistry involved.

The true answer had to wait until the fundamental nature of solar spectroscopy was worked out, the field of nuclear chemistry was developed, and many other technical advancements were made. Though Kirchhoff and other solar spectroscopists had determined that the sun contained most of the same elements that make up the earth's mantle, these researchers were only able to determine what the elements were, not the quantities of each.

The quantitative composition of the sun took a long time to fathom. According to Hufbauer, it was not until 1929 that this part of the puzzle was put into place by a team at the Mount Wilson Observatory, headed by astronomer Henry Norris Russell of Princeton. Russell put forth a convincing set of arguments for a large abundance of hydrogen in the sun (nearly 80%).[11] The abundance of hydrogen on the earth is only about 0.2% so this was an important difference from the composition of the earth. Knowledge of the hydrogen abundance was essential to the understanding of the nuclear theory of the sun's furnace. The nuclear reactions that fuel the sun occur when two hydrogen atoms are fused together at extreme temperature and pressure to form a helium atom. This process releases incredible energy; it is the basic chemistry of our sun and the source of its energy.

Arthur Eddington, the director of Cambridge University's observatory, had already put the gravitational and meteor theories to rest by rigorous theoretical calculations. Furthermore, he had shown that the sun's physical properties could be described in terms of gaseous behavior and that the ideal gas laws applied to the sun as well as other stars. In 1926, Eddington wrote a classic monogram entitled *The Internal Constitution of the Stars,* which synthesized not only the physical data, but also the chemistry of the nuclear reactions that fueled the solar engine.[12] Eddington's work presents a picture of the sun that is very close to what is currently believed. Clearly, the concepts regarding the sun's internal structure have been refined since Young's studies, and the sun's chemical processes are much better understood. Table 7.1 provides some of the information about the sun as science has progressed from Herschel's understanding to that of today. For example, it is now known that the sun is a gaseous sphere—made up mainly of hydrogen and helium—that rotates once in about twenty-six days at its equator and about thirty-two days at its poles.

Table 7.1. Comparison of Solar Data From 1850 to the Present

Sun's property	1849[a]	1896[b]	1990[c,d]
Distance to the Earth (miles)	95,000,000	92,885,000	92,958,257
Diameter of the Sun (miles)	882,000	866,400	864,100
Mass ratio of the Sun to the Earth	354,432	331,000	330,000
Time for one rotation of the Sun	25 days	25 days	26 days at equator 32 days at the poles
Temperature of the Sun's surface	>2,000 K	8 to 10,000 K	5770 K
Age of the Sun (years)	Unknown	Unknown	4.7 Billion
Solar constant	Unknown	3 cal/min/cm^2	2 cal/min/cm^2
Composition	Unknown	Earth-like	78% Hydrogen 20% Helium 2% Other

[a]Herschel, S.J.F.W., *Outlines of Astronomy* (London: Printed for Longman, Brown, Green, and Longmans: 1849)

[b]Young, C.A., *The Sun.* The International Scientific Series (New York: D. Appleton and Company: 1896)

[c]Gribbin, J., *Blinded by the Light: The Secret Life of the Sun* (New York: Harmony Books: 1991)

[d]Lerner, R.G. and G.L. Trigg, ed., *Encyclopedia of Physics.*, 2nd ed. (New York, VCH Publishers, Inc: 1991)

THE CURRENT UNDERSTANDING OF THE SUN

Scientists now know that the sun's energy is fueled by hydrogen. They believe that this nuclear furnace has been generating energy for 4.7 billion years and will continue for billions of years to come. The sun will eventually "burn" all its hydrogen and start another nuclear cycle by burning the helium that was produced from the hydrogen fusion reactions, probably about 1.5 billion years from now. In his book *Blinded by the Light*, John Gribbon, well-known science writer, states that the sun has increased in brightness (and in its energy and heat output) over the millennia; it is now 40% brighter than at its birth. This trend will continue and it will be 15% brighter yet by the time it is six billion years old.[13] This should not worry us unduly, however, as the time humankind has been on earth amounts to only a few seconds on this galactic time scale.

Table 7.2. Nature of the Sun[a]

Major Shell	Thickness (fraction)	Temperature range (K)	Average density (g/cm^3)
Core[b]	0.25	13,000,000 to 15,000,000	89
Radiation zone[c]	0.60	1,500,000 to 13,000,000	10
Convection zone[d]	0.15	5770 to 1,500,000	0.005
Photosphere[e]	0.0007	5770	0.0000002

[a]Adapted from J. A. Eddy, "The Sun," in: *The New Solar System*, J. K. Beatty, B. O'Leary, and A. Chaikin, eds. (Cambridge: Cambridge University Press, 1981)

[b]This is the "core" of the nuclear furnace. It is the only place where the conditions are acceptable for the nuclear fusion reactions to take place. Under these conditions the photons that are produced as products of the nuclear reactions can travel only fractions of a centimeter before undergoing collision with other particles.

[c]In this inferno it takes a photon emitted in the core some 10,000,000 years to reach the surface of the sun. This is boiling action beyond human conception!

[d]The convection zone acts somewhat like water being heated on a stove or the desert air being heated by the noon-day sun. Hot material rises and cool material replaces it at the bottom, soon to be heated and rise as well. This process goes on until the photosphere is reached.

[e]This very thin zone is what is visible to the human eye on earth. The temperature of the photosphere is what is measured and felt on earth. It is so thin that it finally permits the radiation to go freely away from the sun. This light that has taken millions of years to reach the sun's surface can now travel the 93,000,000 miles to the earth in only about eight and a half minutes.

Scientists now believe that there are four zones inside the sun: the core, the radiation zone, the convection zone, and the photosphere; these are described in Table 7.2, which provides some insight into the nature of each.

Even though the sun's core contains only 1.5% of the sun's volume, it holds over half of the sun's mass, or weight. The pressure in the core is 300 billion times that of the earth's atmospheric pressure! The nuclear reactor operates in the core, which is where the fusion of hydrogen takes place, producing the sun's energy. The core is the hottest part of the sun and reaches temperatures of fifteen million degrees. It is also the densest; it is twelve times denser than lead.

During the nuclear fusion reaction that combines two hydrogen atoms to form a helium atom, a photon of radiational energy, otherwise known as electromagnetic radiation, is emitted. This photon, when produced in the core, travels only millimeters before it reacts with a charged particle; the charged particle becomes "excited" and reemits radiation (or another photon) in a random direction. Before the radiation goes almost any distance, it interacts with another charged particle and the process starts once again. This is repeated billions of times until the radiation finally reaches the convection zone. The original photon emission takes millions of years to result in radiation escaping the radiation zone, ten million years on average. The radiation zone is less dense and at its edge cools down to only 1.5 million degrees. At this "cool" temperature the particles no longer fly apart and the energy absorbed by the particles results in heat. The radiation zone makes up the largest volume of the sun and is still ten times more dense than water.

The radiation can now enter the convection zone. In it there is mass flow upward due to convectional heating, just as in a pot of water being heated. If one were able to see the top of the convection zone, it would look a lot like water just starting to boil in the pot. At the top of the convection zone the temperature has cooled even further and is just over 6,000°C (almost 11,000°F), which is the solar temperature measured with earthbound instruments. The pressure and density have also decreased by orders of magnitude. Now at last the radiation can escape from the sun.

The photosphere is the very thin zone where the radiation escapes; it is the part of the sun that is measured to obtain the sun's diameter, and it is the only part of the sun's radiation that can be seen from the earth, except during a total solar eclipse. The sunspots and other signs of the sun's activity that are visible from the earth occur in the photosphere.

Outside the photosphere, there is a relatively thin shell of solar atmosphere called the chromosphere, in which great prominences and solar flares occur. During a total solar eclipse it is possible to see these great gaseous emissions that are always being thrown from the sun, and until scientists were able to send instruments into space it was only during solar eclipses that these observations could be made. Much of the radiation emitted from the chromosphere can only be observed from space because the earth's atmosphere is very effective at absorbing all traces of it.

THE VARIABLE SOLAR CONSTANT

The source of the earth's energy (heat) is the continuous radiation emitted by the sun's photosphere. You will remember that this energy is referred to as the solar constant. Before satellite-borne instruments were sent into space, measurement of the solar constant was inaccurate because it required correcting for the amount of heat that was absorbed by the earth's atmosphere. The earth's atmosphere did not cooperate at all in these measurements because of clouds, moisture variation, and many other factors that interfered. With instruments on satellites outside the atmosphere, scientists can now measure the solar energy flux directly.

Needless to say, the advent of the space age provided an important source of information that an earthbound observer could never obtain. Scientists can now look directly at the sun without the earth's atmospheric filter, which completely absorbs or reflects part of the sun's radiational spectrum. Astronomers cannot see X-rays or most of the ultraviolet spectrum of the sun's radiation from observatories on the earth. Using instruments on satellites, they can look at X-radiation and ultraviolet radiation emitted from the sun as well as obtain a much cleaner view in the visible and infrared regions. With this information they have refined the concept of the sun and believe that they understand it much better than ever before.

The solar constant, being the most important measure of the sun's heat that is transmitted to the earth, was one of the most important solar features to be studied throughout the space program. Astronomer John Eddy of the High Altitude Observatory wrote an excellent monograph for NASA, "A New Sun, The Solar Results from Skylab," in which he described the space-based exploration of the sun.[14] The Orbiting Solar Observatory, OSO 1, was launched in 1962; seven more OSOs were launched in this series, which kept the sun under nearly continuous surveillance for seventeen years. *Skylab*, a manned observatory with eight telescopes, was launched in 1973 and spent nine months in solar observation. In 1974 *Helios A* was launched, followed by *Helios B* in 1976. These spacecraft were placed in solar orbits closer than the orbit of Mercury, the planet closest to the sun. Finally, the Solar Maximum Mission, begun in 1980, observed the sun for nearly ten years with the most sophisticated instruments ever. The results of the space program have made it necessary for the textbooks on the sun to be rewritten.

These extensive satellite experiments have provided a great deal of detailed information about the sun, according to J. A. Eddy, especially about its atmospheric emissions from the ultraviolet and X-ray spectral regions. Scientists have gained knowledge about solar flares, magnetic storms, and the turbulent chromosphere.

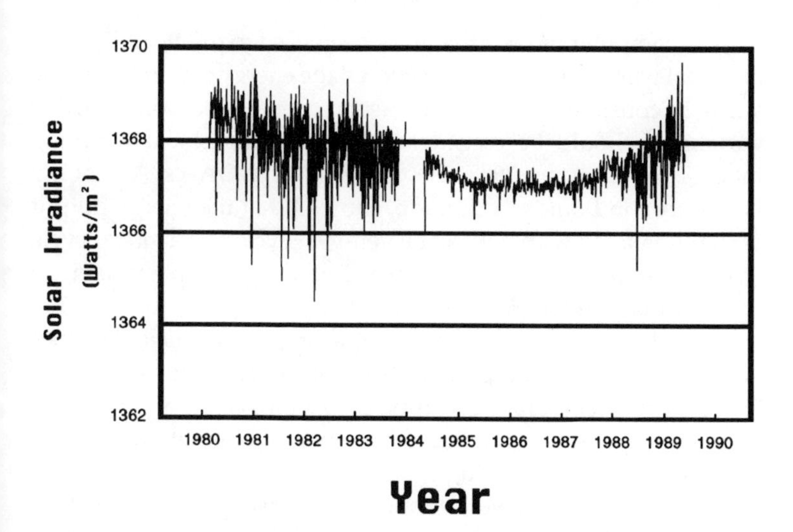

FIGURE 7.3. Variation of the solar constant from 1980 to 1989. These satellite daily mean solar irradiance data were obtained by the ACRIM I instrument on the Solar Maximum Mission. The data were provided by NOAA's National Geophysical Data Center at Boulder, Colorado.

They discovered the solar wind, which constantly sends a flux of particles toward the earth. These are just the tip of the iceberg of important discoveries brought by the space age to solar physicists and astronomers. But most important for our story is the study of the solar constant.

The Active Cavity Radiometer Irradiance Monitor (ACRIM I), which was carried on the Solar Maximum Mission satellite, measured the solar constant about 4,000 times each day for nearly ten years, from 1980 to late 1989. R. C. Wilson and H. S. Hudson published the results from ACRIM I in 1991 in *Nature*.[15] Figure 7.3 provides the results of these measurements. These data provide hard, clear proof of the variability of the sun's energy to the earth.

GLOBAL WARMING MODELS AND THE VARIABLE SOLAR CONSTANT

The computer models that have been used to predict the effects of global warming have all considered the energy coming from the sun to be constant. However, since solar energy is not constant, this variability should be considered in computer global warming models. As early as 1987, George Reid of NOAA's Aeronomy Laboratory in Boulder, Colorado, showed that the energy output of the sun varied with both an eleven-year cycle and a longer cycle of eighty to ninety years.[16] By means of a very simple model, he showed that the temperature variation on earth for the past 100 years could be accounted for by *solar energy variation alone*! He does not insist that solar energy variation is the only cause of variation in the earth's temperature, but he does point out that his simple model more closely explained the temperature variation than did the carbon dioxide doubling experiments.

Reid's conclusions were supported in 1991 by two scientists at the Danish Meteorological Institute, E. Friis-Christensen and K. Lassen, in a study in which they took the length of solar cycles into account, as well as the sunspot data.[17] This study showed a remarkable agreement with the solar cycle length and the temperature variation for the past century.

Only recently have computer climate modelers started to look at the effects of a variable solar flux, or intensity. In most cases these researchers have concluded that the sun's variation is real and should be included in their models; however, they are cautious in attributing too much of the effect to the sun. P. M. Kelly and T. M. L. Wigley of the Climatic Research Unit at the University of East Anglia are two English scientists who have spent considerable research effort on the global temperature of the earth. They incorporated the sun's variation in their computer model and concluded that solar energy variation does play a significant role.[18] It is important to understand that their computer climate model is one of those that has been used in determining that global warming will be caused by an increase in atmospheric carbon dioxide.

Two important climatologists at the University of Illinois, Michael Schlesinger and Navin Ramankutty, performed similar computer modeling experiments including solar energy variation and concluded the following: "Consequently, our results constitute strong evidence, albeit circumstantial, that there have been variations in solar irradiance which have contributed to the observed temperature changes since 1856."[19] They also find that when solar energy variation is included in their model, the effect of doubling the concentration of carbon dioxide is only about half that predicted without including the solar variation in the model. Despite this, these advocates of the greenhouse gas theory of global warming believe that the sun plays only a minor role. The fact remains that the effect of the sun has been ignored in most global warming computer models.

A few astronomers have started to study seriously the variability of the sun with respect to the past temperature history of the earth. The evidence is growing stronger that solar energy variation is an important factor. When the temperature of the earth is compared to the sun's activity, it is important to note that there is a direct correlation between decreased solar activity and the cold temperatures of the seventeenth century. This has been emphasized by Gribbon[20] and by Andrew A. Lacis and Barbara Carlson of NASA's Goddard Institute for Space Studies.[21] You will remember from previous discussions that this cold period was so pronounced that it has been designated as the Little Ice Age.

Solar scientists often study stars that are of approximately the same size, brightness, and age as the sun to help in their understanding of the sun's processes. Sallie Baliunas of the Harvard–Smithsonian Center for Astrophysics and Robert Jastrow of Dartmouth College have observed seventy-four stars of about the same age and mass as our sun. Their studies have been carried out since 1966 for thirteen of the stars and since 1980 for the rest. Baliunas and Jastrow observed that the majority of these sunlike stars have cycles very much like those of our sun.[22] They conclude that our sun may have had longer cycles of solar energy flux that could easily account for the Little Ice Age and similar colder periods of the earth's history. The work of Baliunas and Jastrow indicates that the most important feature is the magnetic activity of the sun, but it is already well established that the magnetic activity correlates with the solar cycle as well.

It is very clear from recent solar studies, especially those observations from space, that the sun, as a source of heat and energy for the earth, is not a constant in its intensity or energy output. Small variations occur on an hourly and daily time frame, but more importantly, larger variations occur with eleven-year and probably longer cycles, which can account for periods of cold such as the Little Ice Age. When the known variation of the sun's energy output is included in the computer models that are used to predict the earth's temperature, the models produce results closer to reality than those models that disregard this variable. There can be no doubt that treatment of the sun as a constant in global climate models is incorrect. Just as Kelly and Wigley[18] and Schlesinger and Ramankutty[19] have shown in their computer climate modeling studies, when the sun's energy variation is included, the predicted effect of increased carbon dioxide is moderated.

REFERENCES

1. J. F. W. Herschel, *Outlines of Astronomy* (London: Printed for Longman, Brown, Green, and Longmans, 1849), p. 227.
2. Ibid., 228.
3. Ibid., 234.
4. Kenneth J. H. Phillips, *Guide to the Sun* (Cambridge: Cambridge University Press, 1992), pp. 11–12.
5. C. A. Young, *The Sun*, The International Scientific Series (New York: D. Appleton and Company, 1896), p. 153.
6. Herschel, 238.
7. K. Hufbauer, *Exploring the Sun: Solar Science since Galileo* (Baltimore: The Johns Hopkins University Press, 1991), p. 55.
8. Phillips, 39.
9. Young, 59, 77–88.
10. Hufbauer, 55–57.
11. Hufbauer, 103–106.
12. A. S. Eddington, *The Internal Constitution of the Stars* (Cambridge: Cambridge University Press, 1926).
13. J. Gribbon, *Blinded by the Light: The Secret Life of the Sun* (New York: Harmony Books, 1991), p. 77.
14. J. A. Eddy, "A New Sun: The Solar Results From Skylab," Rein Ise, ed. (Washington, DC: NASA, 1979).
15. R. C. Wilson and H. S. Hudson, "The Sun's Luminosity over a Complete Solar Cycle," *Nature* 351 (1991):42–44.
16. George C. Reid, "Influence of Solar Variability on Global Sea Surface Temperatures," *Nature* 329 (1987):142–143.
17. E. Friis-Christensen and K. Lassen, "Length of the Solar Cycle: An Indicator of Solar Activity Closely Associated with Climate," *Science* 254 (1991):698–700.
18. P. M. Kelly and T. M. L. Wigley, "Solar Cycle Length, Greenhouse Forcing and Global Climate," *Nature* 360 (1992):328–330.
19. Michael E. Schlesinger and Navin Ramankutty, "Implications for Global Warming of Intercycle Solar Irradiance Variations," *Nature* 360 (1992):330–333.
20. Gribbon, 143.
21. Andrew A. Lacis and Barbara E. Carlson, "Keeping the Sun in Proportion," *Nature* 360 (1992):297.
22. Sallie Baliunas and Robert Jastrow, "Evidence for Long-Term Brightness Changes of Solar-type Stars," *Nature* 348 (1990):520–522.

EIGHT

Water

Water is a principal component of almost every aspect of the global climate, and yet water constitutes the least documented or understood feature of global climate models. Water is ubiquitous and is the only substance to exist naturally in all its phases: solid (ice), liquid (water), and gas (water vapor).

The forms of water affect computer climate models in various and complex ways. Ice and snow are very efficient at reflecting the sun's radiation back into space. Clouds both reflect and absorb the sun's radiation, and they reflect and absorb the earth's radiation as well. On an average day, clouds cover over 60% of the globe, and the cloud cover is ever-changing in location. Water vapor, by contrast, is transparent to most of the sun's radiation but absorbs the earth's radiation, contributing to the greenhouse effect. The oceans absorb the sun's radiation and act as enormous heat sinks that resist temperature change. All these factors greatly complicate the treatment of water in climate models.

THE EARTH'S ENERGY BALANCE

The complex relationships that all forms of water have with the incoming solar radiation and the outgoing radiation emitted by the earth affect the calculation of the earth's temperature in computer climate models. It is necessary for computer modelers to understand these interactions of the sun's warming radiation with each form of water, as well as their subsequent interactions with the earth's radiation. Figure 8.1 illustrates these interactions, describing the energy balance between incoming solar radiation and all of the interactions that occur with it. This simplified picture of energy balance assumes that the incoming radiation from the sun represents 100% of the earth's available energy. We can see that some 6% is backscattered by the atmosphere, 20% is reflected by clouds, and 4% is reflected by the earth's surface (mainly from ice and snow cover) for a total of 30% reflection loss. In addition, some 50% of the solar radiation is absorbed by the earth. Most of this absorption is by the oceans and lakes, which cover over 70% of the earth, but a significant portion is absorbed by plants. This solar radiation is used in the photosynthesis process that causes plants to grow and is therefore the ultimate source of all our food.

The rest of the outgoing radiation identified in Figure 8.1 is that which is emitted by the earth. The earth emits radiation at lower energy than the sun because of the earth's much cooler temperature. The sun radiates energy from its surface at 6,000°C (about 11,000°F), and the earth radiates energy from its surface at about 15°C (59°F).[1] The sun's high-energy radiation occurs at short wavelengths (visible and ultraviolet radiation with wavelengths shorter than about 4 micrometers). Because it is so cool, the earth's radiation is at much longer wavelengths (lower infrared energy). The radiation emissions from both the sun and the earth are illustrated in Figure 1.1 in Chapter 1. The total energy radiated from the earth is minuscule compared with that from the sun.

Remember that the values in Figure 8.1 are all averages. During the course of the year these numbers would vary depending on the location, season, or even whether it is day or night. The Eskimo will never feel the same solar warmth as the Samoan!

If some of the outgoing energy from the earth is captured by atmospheric gases, and if the incoming energy from the sun is greater than the earth's outgoing radiation, the net result will be a warming of the earth (the greenhouse gas effect). As we saw in Chapter 5, the earth's temperature has fluctuated through the

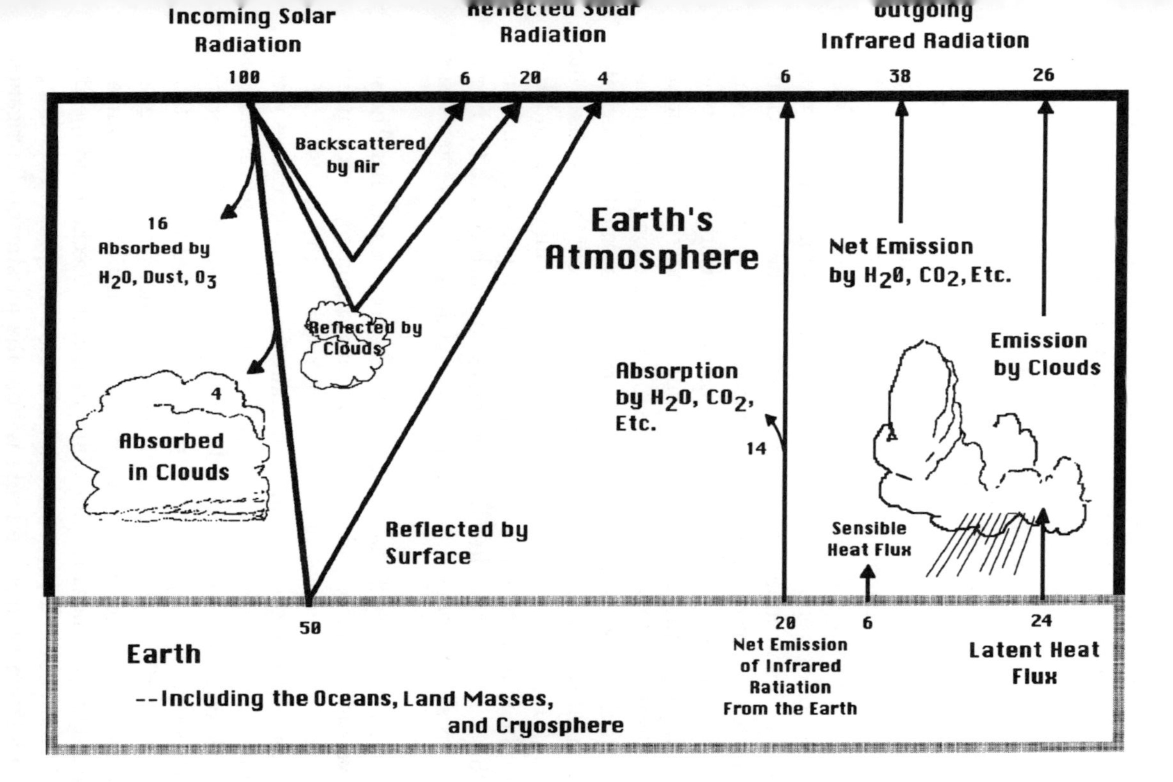

FIGURE 8.1. Average global radiation energy budget. The values are expressed as percentages. This figure was adapted from data in Jose P. Peixoto and Abraham H. Oort, *Physics of Climate* (New York: American Institute of Physics, 1992), p. 94.

centuries, but only very slowly. The earth's resistance to temperature change is regulated by the water, or hydrological, cycle.

THE HYDROLOGICAL CYCLE

The global relationships with respect to all water phases and the processes by which one phase changes to another are described by the global hydrological cycle. But even here intricacies are subtle and the consequences are important for the climate. The water cycle, illustrated in Figure 8.2, seems simple at first glance. Water evaporates from the oceans and lakes, exists for a time as water vapor, then condenses to form clouds over the earth. The clouds create rain, snow, sleet, or hail, and the water returns to the earth or sea. If the precipitation is on land, some is absorbed by the plants and soils, and the runoff goes into lakes and rivers, eventually returning to the sea. Table 8.1 provides information on these fluxes. Note that there is more evaporation from the ocean than precipitation into it; so there is a net transfer of water from the oceans to the land. If precipitation were spread evenly over all of the land area of the world, there would be 29 inches of rain per year. Of course, this is not the case; less than ten inches fall each year in the desert areas and over one hundred inches can fall in the rain forests each year.

The water cycle has a major influence on the weather. The sun warms the earth, and the earth's heat is radiated, which in turn causes large masses of warm air to rise into the atmosphere by means of heat convection. In addition, the sun's heat causes water to evaporate from the oceans and lakes. When water vapor cools and condenses to form clouds, it creates a downdraft of cool air. These columns of rising and falling air, coupled with the rotation of the earth and barometric pressure variations, cause the general atmospheric circulation patterns. This circulation is the source of our winds, which are quite different in both direction and speed at the surface of the earth than higher in the atmosphere. The variable circulation creates the trade winds, the jet stream, hurricanes,

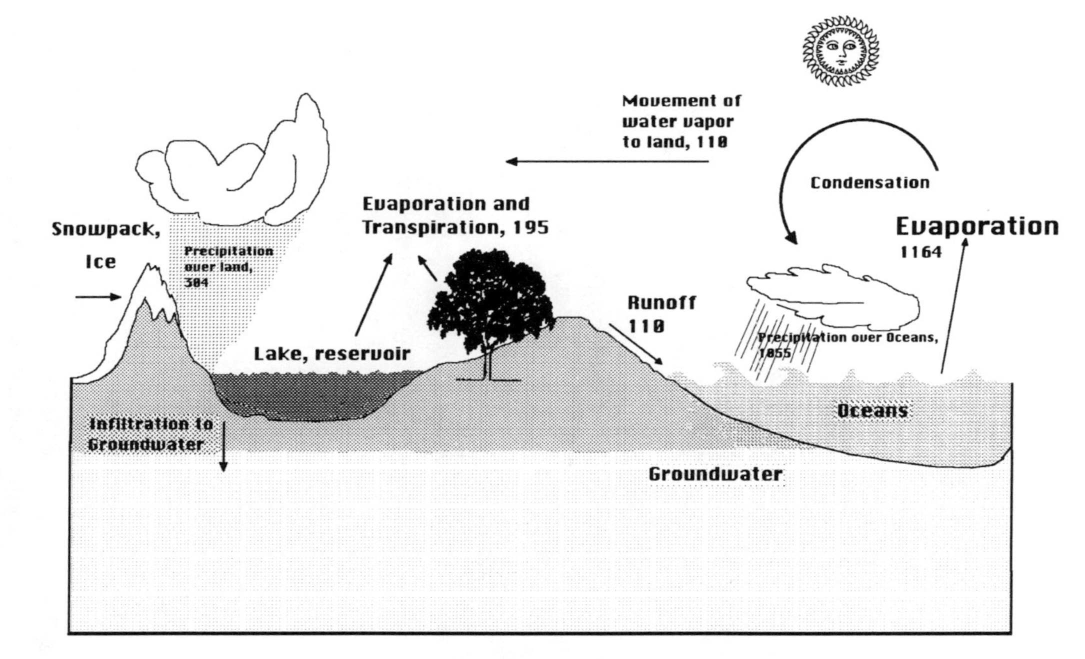

FIGURE 8.2. The water cycle. The numbers represent trillions of liters per day. This figure was adapted from data in Stanley E. Manahan, *Environmental Chemistry*, 6th ed. (Boca Raton, FL: Lewis Publishers, 1994), p. 24.

Table 8.1. The Water Cycle[a]

Process	Flux[b]	
	cm/yr	in/yr
Precipitation on oceans	107	42
Evaporation from oceans	117	46
Precipitation on land	74	29
Evaporation from land	49	19
Runoff from land (river runoff and direct groundwater discharge to the oceans)	25	10

[a]Adapted from Elizabeth Kay Berner and Robert A. Berner, *The Global Water Cycle* (Englewood Cliffs, New Jersey: Prentice-Hall, Inc., 1987) p. 14.

[b]The fluxes in cm/yr and in/yr area based on the following: area of the earth = 510×10^6 km^2; area of the oceans = 362×10^6 km^2; and area of land = 148×10^6 km^2.

typhoons, monsoon rains, tornadoes, fog, dust storms, and all other weather events that we either enjoy or must endure.

This description is an oversimplified version of the atmospheric general circulation model (GCM). Though the general features are easily understood, the patterns are three-dimensional and, because of the many factors that influence them besides the water cycle, not very predictable. The general air circulation of the atmosphere must be modeled very accurately to have any hope of predicting global warming. This is precisely what computer climate modelers strive to do, but they are admittedly a long way from reaching perfection.

In the sections below, we discuss additional complicating factors affecting atmospheric circulation: ocean circulation, ice packs, water vapor, and clouds. It is important to understand the quantities of water found in each of its phases, and Table 8.2 provides an inventory of the water cycle. It is impressive to realize that over 97% of the earth's water is in the oceans, and only 0.001% is in the atmosphere. Only about 2.74% of all the water (rivers and lakes, ice in all forms, and groundwater) on earth is actually in a drinkable form.

Table 8.2. Water/Snow/Ice Inventory Information[a]

Forms of Water	Comments	Volume ($\times 10^6$ km^3)[b]	Percentage of the whole
All forms	Total on earth	1408.7	100.0
Sea water		1370.0	97.25
Rivers & Lakes		0.127	0.009
Ground Water	Shallow and Deep	9.5	0.68
Water vapor	Atmospheric	0.013	0.001
Ice	All forms	29.0	2.05

[a]Adapted from Elizabeth Kay Berner and Robert A. Berner, *The Global Water Cycle* (Englewood Cliffs, New Jersey: Prentice-Hall, Inc., 1987) p. 13.

[b]A km is a kilometer, a 1,000 meters; a km^3 is a cubic kilometer, a volume equal to a cube with 1,000 meters on each side. A km^3 would hold about 264×10^9 gallons, which is 264 billion gallons.

WATER VAPOR

The atmospheric gas most effective at trapping the earth's radiation is water vapor, which accounts for about 60–70% of the greenhouse effect, according to the International Panel on Climate Change (IPCC).[2] The atmospheric concentration of water vapor varies significantly on a regional basis and with altitude. Overall, however, water vapor is present at much greater concentrations than carbon dioxide or any of the other greenhouse gases. Water vapor generally ranges between 1% and 3% of all the gases in the atmosphere, but it can be as low as 0.1% and as high as 5%, according to Professor S. Manahan.[3] Relative humidity is a measure of the amount of water vapor in the air relative to the maximum amount the air can hold at a given temperature; the higher the temperature, the more water vapor the air can hold. This is the reason that when the relative humidity is 80% or 90% in the deep South, it is really steamy, but in the cool mountains high humidity doesn't seem that wet at all.

The actual humidity at any place and time depends on many factors (cloud cover, barometric pressure, winds, moisture content of soils, etc.) and is not very predictable. The relative humidity can vary from a small percentage to 100% during the course of a day. This short-term regional variation is one of the main reasons why computer climate models cannot predict short-term weather changes, much less longer-term changes.

The issue of water vapor and the lack of knowledge of its role in global computer models prompted David Starr and Harvey Melfi of NASA's Goddard Space Flight Center to organize a workshop on "The Role of Water Vapor in Climate" in 1990. In the foreword of their subsequent report they point out this lack of understanding:

> . . . it became clear that present water vapor measurements were inadequate to define the state of the atmosphere on almost any scale; and furthermore, that GCMs [general circulation models of the atmosphere] . . . lacked the ability to realistically incorporate real moisture data. . . . Our lack of knowledge concerning atmospheric water vapor is all the more striking considering the dominant impact moisture processes have in the Earth's weather and climate.[4]

A goal of the NASA workshop was to assess the scientific state of the knowledge of water vapor in the air and the ability to measure it, and to determine what research should be initiated to improve the situation. They acknowledged the importance of atmospheric water vapor in the following fundamental respects:

> 1) Water is the principal medium for direct energy exchange among the major components of the Earth System—atmosphere, hydrosphere, cryosphere, biosphere and lithosphere . . .
>
> 2) Water vapor is *the* predominant greenhouse gas and plays a crucial radiative role in the global climate system . . .
>
> 3) Water vapor is an essential ingredient in many atmospheric processes which are intimately involved in determining specific realizations of climate variations, especially on regional scales . . .[5]

All these statements were endorsed by the sixty-two eminent climate scientists who represented most of the government research organizations and universities doing state-of-the-art climate model research in the United States. Understanding the role of water vapor in the atmosphere is at the very core of understanding the climate.

CLOUDS

The role of clouds in global climate is even more complex than water vapor alone. Clouds have a significant influence on climate. They can consist of all three water phases—ice, liquid, and vapor—as well as particulates and trace gases. There are many types of clouds, such as the wispy cirrus clouds high in the troposphere and the cumulonimbus thunderheads formed near the earth. There are many other types of clouds in every part of the troposphere and even some in the stratosphere. Different types of clouds form at different altitudes and under different conditions. They affect the earth beneath in a variety of ways; some cause cooling and some cause warming. Clouds cover over 60% of the earth on any given day, but the effects of each type cannot simply be averaged because of the way the various types interact to either cool or warm the earth.

The dynamics of cloud formation and subsequent disappearance are chaotic. The temperature of the earth would be far warmer if there were no clouds; it is generally agreed that clouds cause a cooling effect, and yet climate models often associate an increase in cloud cover with increasing temperature. Let us look at some dominant climate interactions caused by clouds:

Cooling effects:

- Clouds reflect or absorb radiation from the sun (thus the sun's energy never reaches the earth; this affects the land, oceans, and ice and snow cover).
- Clouds cool the surrounding air through evaporation of water droplets (evaporative cooling).
- Clouds transfer water to the earth by means of precipitation (rain, snow, sleet, or hail).

Heating effects:

- Clouds heat the surrounding air by transferring their stored heat (if the temperature of the cloud is warmer than the temperature of the surrounding air—like a space heater).

- Clouds reflect or absorb radiation from the earth (thus the atmosphere traps the earth's radiation—the greenhouse effect).
- Clouds radiate energy according to their temperature.

It is easy to see why clouds cause climate modelers fits. If any of the various interactions of clouds are introduced into the models in an exaggerated way, the wrong effect may be emphasized and an erroneous prediction obtained.

As mentioned, the effects of clouds may vary, depending on altitude and type of cloud formation. Another complicating factor is the fact that clouds often form in layers or overlap, another interaction that is difficult to model and that can result in erroneous predictions as well.

Studies of the cloud cover from satellites have shown that, overall, clouds cause cooling, but because cloud dynamics are chaotic and sensitive to turbulent winds, they are still poorly understood. Andrew Revkin, a science writer for *Discover*, reported on a climatology conference with a focus on clouds. In his article, he quotes Bruce Barkstrom of NASA's Langley Research Center as follows: "Clouds form quickly. They're sensitive to turbulence, which we don't understand at all. We'll be in the next century before we really understand what clouds are doing to Earth's climate."[6] Not only do clouds form quickly and chaotically, they are often much smaller than the resolution of the models. Remember, the area resolution of the models is generally quite large, about the size of New Mexico. When events are smaller than the resolution of the computer model, the model cannot explicitly account for their effects.

It is fair to say that climate modelers have yet to account adequately for the many complex and often opposite cloud effects in their computer models. In early models, scientists simply held the cloud cover as a constant so they would not have to further account for it at all! The early models predicted greater increases in global warming than more recent ones, which are more sophisticated in their cloud treatment. It appears that the more realistic the global climate models become, the smaller the temperature change they predict!

OCEANS

If water vapor in the atmosphere and the various cloud effects complicate the understanding of climate, the role of oceans stands as Goliath to David by comparison. In Chapter 5 we discussed the fact that scientists had an inadequate data base from which to determine the earth's temperature. Even less information is available with respect to the ocean's circulation. Carl Wunsch of the University of Cambridge states the situation as follows: "Existing knowledge of the ocean circulation is based upon a combination of fragmentary . . . observations (hydrography), and a number of highly plausible theoretical ideas which have rarely been tested directly."[7] Whereas many research organizations are making great strides toward a better understanding of ocean circulation and systematic measurement of its important parameters, baseline information is lacking. There have been few long-term programs to obtain the important parameters of temperature, current speed and direction, and salinity, among others, relative to depth in the oceans. Not only is the oceanic data base fragmentary, there is very little of it.

The fact that the oceans absorb about half of the solar energy that reaches the earth is essential to the behavior of the global climate. This energy is not evenly distributed, most of it being absorbed within the area between 30° N and 30° S latitudes. The oceans generally heat up slowly during the summers and release heat to the atmosphere as they cool down during the winters; thus, they are gigantic heat sinks that moderate the earth's temperature. In addition, some of the warm water near the equator moves toward the poles and provides one of the sources for ocean currents. An averaged energy distribution for both landmasses and the oceans is illustrated in Figure 8.3. Note that oceans and continents do not receive an equal distribution of radiation in the Northern and Southern Hemispheres.

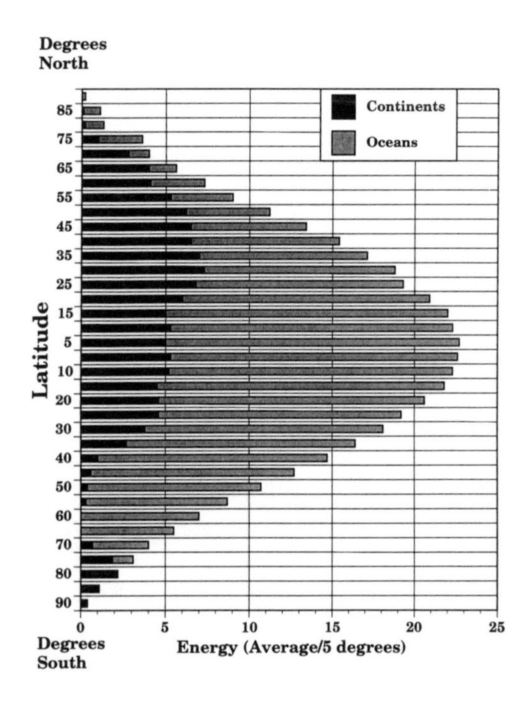

FIGURE 8.3. Average solar energy striking the earth's oceans and continents as a function of latitude. The units are $10^{22} \times$ joules per 5° latitude. This figure was adapted from data in J. D. Woods, "The Upper Ocean and Air–Sea Interaction in Global Climate," in *The Global Climate*, ed. John T. Houghton (Cambridge: Cambridge University Press, 1984), p. 142.

If this uneven distribution were the only complicating factor, computer models might easily deal with the situation, but it is further complicated by the depth to which the sun's energy affects the ocean waters. The solar warming reaches to about 100 meters (over 300 feet) deep into the ocean on a cloudless day. Depending on the depth and the latitude, several things may happen to the warmed water:

- It may mix with deeper, colder ocean water, but only during the winter.
- It may mix with deeper, colder ocean water all year.
- It may never mix with deeper, colder ocean water.

This mixing also helps to drive ocean currents. Like the atmosphere, the ocean has a circulation, driven by the movement of the huge masses of warmer and cooler water.

Cold water is more dense than warm water. Pure water is most dense when its temperature is 4°C (39°F). The cold water sinks toward the bottom of the ocean, and warm water rises toward the surface. These movements help to create the major ocean currents. The currents are further affected by the topography of the ocean floor, which influences areas of upwelling and downwelling.

Ocean circulation is further complicated by the saltiness, or salinity, of the water, which affects its density and temperature. There are certain salinity and temperature ranges that create layers in the ocean. (If you have ever gone swimming in a lake, you have undoubtedly felt a colder layer of water a few feet below the surface.) These layers are so pronounced in the oceans that oceanographers give them names. Pearn Niiler of Scripps Institution of Oceanography at La Jolla, California, describes five layers in the Atlantic Ocean, emanating from distinct sources, as follows:

- Antarctic Bottom Water—The most dense and deepest is below 3500 meters, with a source on the Antarctic Continental Shelf.
- North Atlantic Deep Water—lies between 4000 meters and 2000 meters and appears as emanating from north of the Gulf Stream. It can be traced to the Norwegian and Labrador Seas.
- Mediterranean Water—Another salt tongue at the 1500 meter level that can be traced eastward to the Mediterranean Sea.
- Antarctic Intermediate Water—emanates from the Circumpolar Current area at 800–1000 meters.
- Sub-Tropical Mode Water—the most salty is a lens on the surface caused ostensibly by high evaporation under the Trade Winds.[8]

Figure 8.4 provides a sketch of these water masses with their relative positions in the North Atlantic at 28° N latitude.

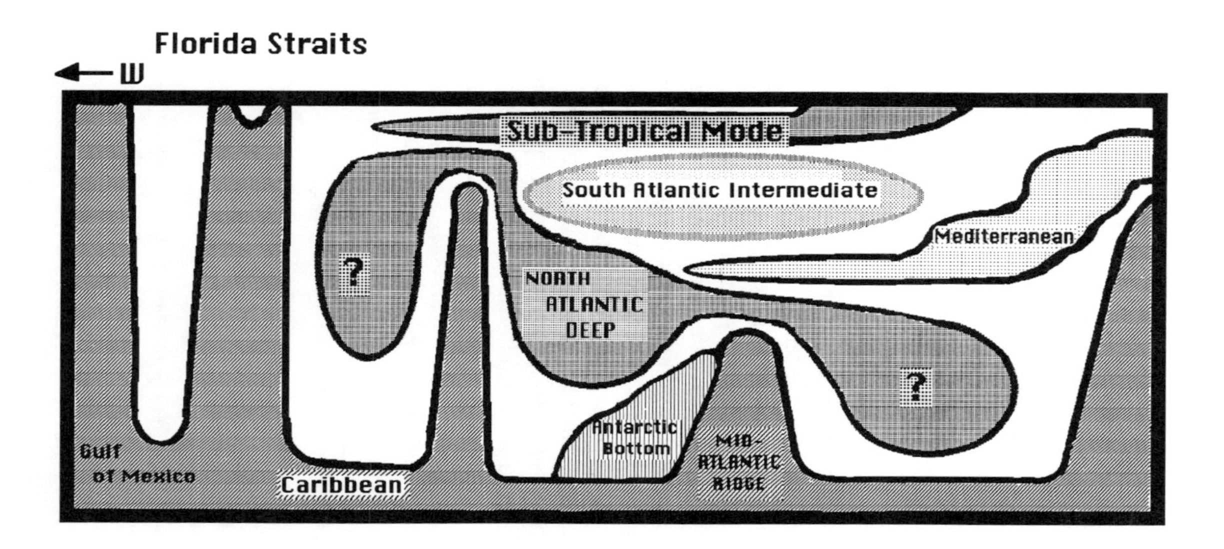

FIGURE 8.4. Ocean water "types" in the North Atlantic at 28° N latitude. This figure was adapted from P. P. Niiler, "The Observational Basis for Large Scale Circulation," in *General Circulation of the Ocean*, ed. Henry D. I. Abarbanel and W. R. Young (New York: Springer-Verlag, 1987), p. 45.

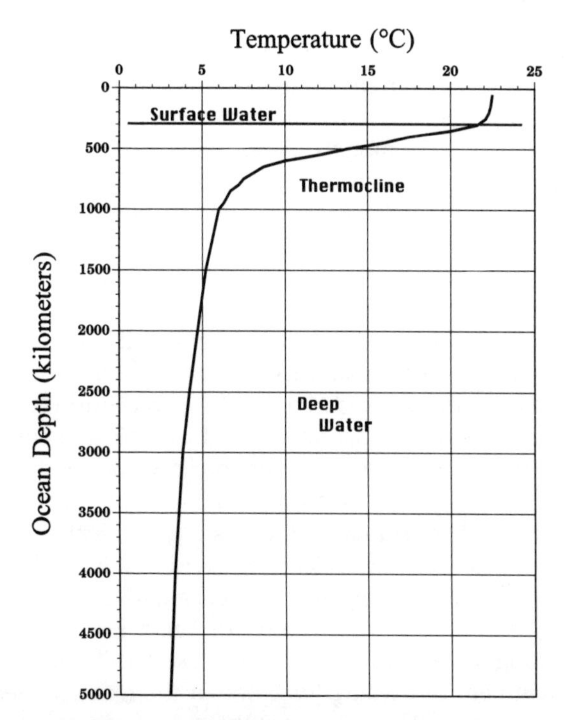

FIGURE 8.5. Typical ocean temperature as a function of depth. This figure was adapted from Elizabeth Kay Berner and Robert A. Berner, *The Global Water Cycle* (Englewood Cliffs, NJ: Prentice-Hall, 1987), p. 30.

As stated, the different layers of water have definite salinity, density, and temperature relationships. Figure 8.5 illustrates a typical curve of ocean temperature relative to depth. The temperature is generally about 20°C (68°F) from the surface down to about 100 meters where solar radiation penetrates, then it decreases sharply to about 4°C (39°F) at a depth of about one kilometer. Oceanographers refer to the steep change in temperature as the "thermocline." The variation in this curve throughout the oceans is extremely important to the ocean circulation, and the ability to model the thermocline is essential to the ocean circulation models. It is this information on temperature and current velocity as a function of depth that is lacking from the computer climate modelers' data base. This is the major reason that the climate modelers must make do with "hokey" ocean models, as described by Stephen Schneider, when commenting on the ocean model used by Jim Hansen. (Hansen, you will recall, is the scientist who testified before Congress that "global warming had begun.")

There are, very broadly, two types of ocean current flow, laminar and turbulent. Laminar flow is smooth and orderly, like a deep running river, flowing slowly and peacefully. It can be described by means of simple mathematical equations. By contrast, turbulent flow is like a churning rapid in a narrow gorge. Turbulent flow is nonlinear and therefore chaotic, with whirlpools and eddies, and it is characterized by a great variety of flow speeds. By its very nature this nonlinear, chaotic process is nearly impossible to model mathematically. Unfortunately, there is a great deal of turbulent flow within ocean currents. Scientists David Neelin, of the University of California at Los Angeles, and Jochem Marotzke, of Massachusetts Institute of Technology, in their article in *Science*, "Representing Ocean Eddies in Climate Models," call the mathematical treatment of ocean eddies one of the main challenges in modeling the ocean. They point out that ocean eddies are small in scale when compared to the grid size used in ocean models, which means that they must be included in the models by indirect mathematical expressions (the "parameterizations" discussed in Chapter 4) rather than explicit, or fundamental, equations.[9] This has not been done very well in the past. The problem is analogous to the atmospheric problem of clouds, in which clouds are often small with respect to the size of the model grid.

Processes in the oceans occur at a much slower pace than on land and in the atmosphere. Water vapor lasts only about ten days in the atmosphere before it becomes part of a cloud or condenses back into liquid. Water in the soils lasts maybe a year. By contrast, the residence time of water in the oceans is roughly 3,000 years.[10] Given these tremendous time differences, you can see how difficult it must be to couple ocean models with atmospheric climate models. There are ocean effects that are very slow and cause climate effects over decades. In early models these ocean effects were ignored by climate modelers or greatly oversimplified. Only recently have computer modeling experiments been reasonably successful using realistic ocean models. The computer time required to incorporate a good ocean model into an atmospheric circulation model is enormous. The ocean models require at least as much computer time as atmospheric general circulation models, and when coupled together the resulting time requirements multiply rather than add. This could add weeks to the time required for a single experiment of a century-long simulation.

THE CRYOSPHERE (ICE AND SNOW)

Because of their high reflectivity, the ice sheets of Antarctica and Greenland, the mountain glaciers, and winter snow cover are very important in global climate models. The earth's albedo is the total reflection of the sun's radiation back into space. The average albedo for the earth and its atmosphere is 30%, as was illustrated in Figure 8.1. However, different substances vary markedly in albedo. Table 8.3 gives the albedo values for various types of surfaces. Ice, snow, and white clouds can reflect 70–80% of the sun's radiation, whereas oceans and land reflect very little, only 2–6%, absorbing most of the solar radiation. The variation in albedo numbers reflects the fact that sunlight reflection depends on the angle of the sun with respect to the earth's surface, or the angle of incidence.

Important statistics about the water–ice inventory are presented in Table 8.2 and Table 8.4. Of particular interest is the overwhelming amount of water that rests in the oceans, the almost trivial amount in the atmosphere, and the great variation of snow and ice coverage over both the land and sea from winter maxima to summer minima.

The significant variations in the amount of snow and ice from season to season are illustrated in Table 8.4. At the peak of winter snow coverage in the Northern Hemisphere, about 50% of the land surface is covered and 10% of the sea surface is covered. In the

Table 8.3. Albedo Values[a] for Various Earth Surfaces[b]

Surface	Albedo (%)[c]
Ice or fresh snow	75–95
Old snow	40–60
Sand or deserts	18–35
Grasslands or green crops	15–25
Forests	14–20
Dense or evergreen forests	7–15
Cities	14–18
Plowed fields	10–15
Asphalt	about 8
Water (lakes and oceans)	2–6
Clouds (depending on type)	20–70

[a]The albedo is the amount of the sun's radiation that is reflected back into space and therefore unavailable for warming the earth.

[b]Adapted from Jose P. Peixoto and Abraham H. Oort, *Physics of Climate* (New York: American Institute of Physics, 1992) pp. 103–104; and Stanley E. Manahan, *Environmental Chemistry*, 6th Ed. (Boca Raton, Lewis Publishers: 1994) p. 434.

[c]The values of albedo must be given in ranges because the reflection of the sun varies depending on its angle with respect to the earth's surface, which will change during the day and from summer to winter.

summer only about 2% of the land and 5% of the ocean are covered. Remember that it is summer in the Southern Hemisphere when it is winter in the Northern Hemisphere and vice versa. These major seasonal variations must be accurately included in the computer models for accurate results. The oceans, including the ice–ocean–atmosphere interface, pose major obstacles to the validity of global computer models.

An additional important factor in the water cycle is the extent and depth of the permafrost. Change in the permafrost is measured in centuries rather than decades. The coverage of the permafrost

Table 8.4. Winter and Summer % Snow/Ice Coverage[a]

	Northern hemisphere		Southern hemisphere	
	land	ocean	land	ocean
Summer	2.0%	5.2%	28.6%	1.2%
Winter	49.0%	9.7%	>28.6%	9.7%
	Land + Ocean		Land + Ocean	
Summer	4%		6.5%	
Winter	25%		13.5%	

[a]Adapted from N. Untersteiner, "The Cryosphere," in: *The Global Climate*, John T. Houghton, ed. (Cambridge: Cambridge University Press, 1984) p. 124.

regulates the amount of water evaporated into the atmosphere over great portions of the higher latitudes (near the poles) in both hemispheres.

The ice sheets are also important in that they contain about 80% of the fresh water on earth. If the ice sheets start to melt, they will add fresh water to the oceans, which will affect the salinity of the oceans around the ice sheets. This imbalance in salinity will cause changes in the ocean–ice–atmosphere relationship, which will in turn create long-term (hundreds of years) changes in ocean circulation.

PREDICTED OCEAN RISE WITH GLOBAL WARMING

There have been dire predictions of a large sea level rise associated with the doubling of carbon dioxide in the popular myth of global warming. Though these predictions are mostly exaggerations, the IPCC report summarized estimates from different researchers of upward to 3.65 meters (12 feet).[11] However, the IPCC report proposed more modest values of from 9–29 centimeters (3.5–11.4 inches) for probable ocean rise due to global warming.[12]

Table 8.5. Predicted Contribution to Sea Level Rise (1985–2030)[a]

Source	Low estimate (cm)[b]	High estimate (cm)	Best estimate (cm)
Thermal expansion	6.8	14.9	10.1
Mountain glaciers	2.3	10.3	7.0
Greenland ice sheet	0.5	3.7	1.8
Antarctica ice sheet	–0.8	0.0	–0.6
Total	8.7	28.9	18.3

[a]Adapted from J. T. Houghton, G. J. Jenkins, and J. J. Ephraums eds., *Climate Change: The IPCC Scientific Assessment* (Cambridge: Cambridge University Press, 1990) p. 276.
[b]All sea level rise values are in centimeters (one inch = 2.54 centimeters).

Table 8.5 shows the IPCC estimates of the various factors that may contribute to this rise, including thermal expansion, melting glaciers, and breakup of ice sheets.

If oceans are going to rise significantly because of global warming, which was a major theme of environmentalists at the end of the 1980s, the bulk of the increase would have to be due to breakup and melting of the ice sheets of Antarctica and Greenland. These ice sheets are much more important to sea level rise than the mountain glaciers in northern Europe and America; if all mountain glaciers were to melt, the sea level rise would only be about 25–45 centimeters (10–18 inches), according to the IPCC.[13] The consensus of the scientists in the IPCC is that the ice sheets will not break up or melt in the near future; scientific assessment of potential ice sheet melting indicates that this would take centuries to happen.

Recent studies of the variation in ice and snow accumulations over the Greenland ice sheet for the past 18,000 years, carried out by Professor W. R. Kapsner and a research group from Pennsylvania State University and the University of Washington, have found that variation in storm tracks, not variation in temperature, is the important factor for determining precipitation variability.[14] In discussing these results, David Bromwich, of Byrd Polar Research Center at Ohio State University, states:

> The realization that precipitation variability over the polar ice sheets depends on far more than just temperature changes means that predicting the effects of "global warming" on sea level is not as straightforward as attempted in the report by the Intergovernmental Panel on Climate Change.[15]

Bromwich goes on to discuss the need for better spatial resolution in GCMs before they can adequately assess these storm track versus temperature effects—all of which indicates yet another area in which computer climate models do not include an adequate mathematical model of the climate.

EL NIÑO

El Niño is a major climate event that occurs every three to five years; it affects climate all over the world, causing both flooding and drought. It has no known relationship to greenhouse gases or global warming, but it has profound global effects. If computer models are to be used for predicting climate 100 years in the future, they must first be able to predict successfully such major oceanic events as El Niño.

El Niño is a natural, quasi-periodic change in the sea-surface temperature of the ocean currents off the coast of Ecuador and northern Peru in South America. Sea-surface temperature can be abnormally warm or cool. If it is warm, it is called El Niño, and if cool, La Niña. El Niño generally begins near Christmas and hence the name, which is Spanish for little boy, or "the Christ Child."

During El Niño, there is a major warming of the ocean surface temperature near South America, with an associated change in atmospheric pressures and trade winds. The pressure change is a "seesaw" effect between the southeastern tropical Pacific and the Australian–Indonesian region. When one of these areas has abnormally high pressure, the other has abnormally low pressure, and vice versa. The pressure phenomenon was observed independently and termed the Southern Oscillation. Since this connection between the Southern Oscillation and El Niño has been recognized, the acronym ENSO (El Niño–Southern Oscillation) has been used, according to Henry Diaz and Vera Markgraf in their recent monograph *El Niño, Historical and Paleoclimatic Aspects of the Southern Oscillation.*[16] Their book discusses many aspects of the phenomenon, which has been studied by many climatologists during the past two decades.

It is now accepted that this major and natural quasi-periodic event affects the global climate in a real and significant way.[17] There are areas of increased drought and other areas of increased rainfall at far-reaching places on the globe. The drought–wet conditions in Southern California, the southeastern United States, and central Africa all seem to have connection to ENSO. For example, a group of scientists from Columbia University and Zimbabwe recently observed a relationship between the maize yield in Zimbabwe and ENSO.[18] When the sea-surface temperature off the South American coast increases, maize yields increase halfway around the world in Africa!

Scientists do not fully understand the causes of El Niño. There are computer models that have been somewhat successful in predicting El Niño events. However, according to Richard Kerr, writer for *Science*, the computer models have been inconsistent, for example, missing the prolonged El Niño beginning in 1993.[19] The GCMs used to predict global warming are essentially the same computer models used to predict El Niño, but they have not incorporated the El Niño components when predicting global warming due to increasing greenhouse gases.

All aspects of the global water cycle discussed in this chapter—water vapor, clouds, the oceans, and ice—constitute the most important components of computer climate models. All processes in the ocean are affected by the amount of cloud cover, the diurnal variation in the sun's radiation, the earth's rotation, the pull of the moon's gravity, which causes the tides, and the seasonal effect of the earth's orbit around the sun. It seems clear that these hydrological effects are the least well known features of the computer models.

REFERENCES

1. Jose P. Peixoto and Abraham H. Oort, *Physics of Climate* (New York: American Institute of Physics, 1992), pp. 93, 118.
2. J. T. Houghton, T. J. Jenkins, and J. J. Ephraums, eds., *Climate Change: The IPCC Scientific Assessment* (Cambridge: Cambridge University Press, 1990), p. 48.
3. Stanley E. Manahan, *Environmental Chemistry*, 6th ed. (Roca Raton: Lewis Publishers, 1994), p. 300.
4. D. O'C. Starr and S. H. Melfi, eds., *The Role of Water Vapor in Climate*, NASA Conference Publication 3120 (Washington, DC: National Aeronautics and Space Administration, 1991), p. iii.
5. Ibid., ix.
6. Andrew C. Revkin, "Clouds in the Greenhouse," *Discover* June 1989:24.
7. Carl Wunsch, "The Ocean Circulation in Climate," in *The Global Climate*, ed. John T. Houghton (Cambridge: Cambridge University Press, 1984), p. 189.
8. P. P. Niiler, "The Observational Basis for Large Scale Circulation," in *General Circulation of the Ocean*, ed. Henry D. I. Abarbanel and W. R. Young (New York: Springer-Verlag, 1987), p. 43.
9. J. David Neelin and Jochem Marotzke, "Representing Ocean Eddies in Climate Models," *Science* 264 (1994):1099–1100.
10. Peixoto and Oort, 17–20.
11. Houghton, Jenkins, and Ephraums, 275.
12. Ibid., 276.
13. Ibid., 267.
14. W. R. Kapsner, R. B. Alley, C. A. Schuman, S. Anandakrishnan, and P. M. Grootes, "Dominant Influence of Atmospheric Circulation on Snow Accumulation in Greenland over the Past 18,000 Years," *Nature* 373 (1995):52–54.
15. David Bromwich, "Ice Sheets and Sea Level," *Nature* 373 (1995):18–19.
16. Henry F. Diaz and Vera Markgraf, eds., *El Niño, Historical and Paleoclimatic Aspects of the Southern Oscillation* (Cambridge: Cambridge University Press, 1992), p. 1.
17. W. J. Burroughs, *Weather Cycles, Real or Imaginary?* (Cambridge: Cambridge University Press, 1992), pp. 107–117.
18. Mark A. Cane, Gidon Eshel, and R. W. Buckland, "Forecasting Zimbabwean Maize Yield Using Eastern Equatorial Pacific Sea Surface Temperature," *Nature* (1994):204–205.
19. Richard Kerr, "El Niño Metamorphosis Throws Forecasters," *Science* 262 (1993):656–657.

NINE

Aerosols

In the last three chapters, we discussed the possible climate effects of various factors, including greenhouse gases, variation in solar radiation and the earth's orbit, and the hydrological cycle. A powerful influence that has been largely ignored by climate modelers is that of atmospheric particles, or aerosols. The term *aerosol* refers, in the scientific sense, to very small particulate matter (liquid or solid), so small that it floats about on air currents for days, weeks, or even years at a time. Aerosols in this context are not to be confused with the spray-can aerosols in common household use. Household aerosols are usually solutions or powders sprayed by means of a pressurized gas. These sprays, composed of coarse particles that last for a short time before falling to the ground, are not a source for the atmospheric aerosols discussed here. (However, the gases used as propellants in these aerosol cans are often classified as greenhouse gases.) The aerosols we discuss are from several sources, primarily volcanic eruptions, ocean dynamics, and human activities.

In early computer climate models, aerosols were given little or no attention because they were believed to have a minor or even negligible effect on global climate. The 1992 Intergovernmental Panel on Climate Change (IPCC) report reflects a major change in this attitude, giving greater attention to the cooling effects of aerosols because of both their direct reflection of solar radiation and their indirect participation in cloud formation. In fact, the importance of the aerosol cooling effect is recognized to be nearly equal in magnitude, but opposite, to the warming effect caused by the increase of greenhouse gases.[1] This means that in the two years after the first IPCC report, the scientists on the panel found a global cooling process that equals the dreaded global warming!

The suddenness of this discovery of climatic effect due to aerosols is surprising because the phenomenon has been well known, if not well understood, since the early part of this century. In 1991 research groups from the Department of Atmospheric Sciences at the University of Washington and the Department of Meteorology at Stockholm University published a paper summarizing the early papers that pointed to a cooling effect caused by aerosols.[2] In fact, one of these early papers was published in 1971 by Stephen Schneider, noted global warming scientist. Schneider's paper predicted global cooling based on anthropogenic aerosol effects.[3] In Chapter 2 we discussed Schneider's switch in position from global cooling prediction to a belief in global warming.

AEROSOL COMPOSITION

What are the qualities of these aerosols thought to cause global cooling? They are microscopic particles, or droplets, that exist for a relatively long time in the atmosphere because of their extremely small size. Formed from a multitude of sources, both natural and man-made, they have a varied and complex chemical makeup. Aerosols include silica dust particles from the desert, sodium chloride salt particles from the ocean, sulfuric acid droplets from burning of sulfur-bearing coal or oil, and sulfuric acid droplets originating from volcanic eruptions. Over the continents, a significant portion of particulate matter is formed by decaying biological matter, such as insects, bacteria, and pollen. And soot particles are formed from carbon wherever carbon-containing fuels such as coal, oil, wood, or paper are burned.

Each type of aerosol/particulate has its own interaction with incoming solar radiation and outgoing global radiation. These interactions depend on the aerosol's chemical composition, its particle number density, and the variation in size of its particles. The type and nature of the particles, as well as their location (height) in the atmosphere, are important to their atmospheric effects.

Professor G. M. Krekov of the USSR Academy of Sciences defines five atmospheric altitude ranges used in developing mathematical models of aerosol behavior:[4]

1. An earth surface boundary layer from the earth's surface to 2 kilometers (1.2 miles).
2. A turbulent mixing layer from 2–4 kilometers (1.2–2.5 miles).
3. The tropospheric background layer from the top of the turbulent mixing layer to the top of the troposphere, 10–12 kilometers (6–7.5 miles).
4. The stratosphere from about 12–30 kilometers (7.5 to ≈19 miles).
5. The atmosphere above 30 kilometers (≈19 miles).

Aerosols range in size from about 0.001 to 10 micrometers. (A micrometer is equal to one millionth of a meter.) To give you an idea of the sizes involved, visible cigarette smoke particles are about one to ten micrometers in size. The number density of particles at various locations in three different size ranges is given in Table 9.1. Remember that this number density will vary with time, and these numbers are only averages. The actual number will fluctuate, depending on size distribution, winds, relative humidity, altitude, location, and a number of other factors. In general, the smaller the particles, the longer they float around in the atmosphere before they become part of a cloud or a raindrop or, in their random meandering, fall to the ground or into the ocean. The smallest particles are so small that they would not be visible under a conventional optical microscope. Particles up to about 0.1 micrometer last much longer than do larger ones; however, particles in the range of 0.1 to 1 micrometer are important as well. Particles larger than about 1 micrometer last only a short time in the atmosphere and are not considered important to long-term climate studies.

Table 9.1. Particle Number Concentration for Three Size Ranges and Various Locations[a]

Location	Particle number concentration (number/cm^3)[b]		
Size range and name	0.001–0.01μm[c] nucleation[d]	0.01–0.1μm accumulation[e]	>0.1μm coarse[f]
Polar	16.3	5.8	0.015
Background	10.3	41.9	0.077
Ocean	125	52.6	0.28
Remote continental	1260	81.1	0.44
Desert dust storm	1260	185	14.7
Rural	8610	1.7	0.34
Urban	135,000	1410	0.76
Stratosphere (20km)	0.4	4.1	0.018

[a]Adapted from G. M. Krekov, "Models of Atmospheric Aerosols," in: *Aerosol Effects on Climate*, S. Gerard Jennings, ed. (Tucson, AZ: The University of Arizona Press, 1993) pp. 13–15.

[b]The number of particles per cm^3 is the number of particles in a cube with each side equal to one centimeter in length. If there were 1 particle per cm^3 there would be a million particles (1,000,000 particles) in a cubic meter.

[c]A μm is a micrometer (1×10^{-6} meters, or 0.000039 inches).

[d]Nucleation-sized particles are small enough to act as cloud condensation nuclei (CNN); this is the most important size range because they will last a long time in the atmosphere and affect the weather by both reflecting radiation and forming clouds.

[e]Accumulation-sized particles are considered too large to form clouds but will reflect sunlight.

[f]Course particles are so large that they will last for only a short time in the atmosphere and are not considered to have a long term effect on the climate.

AEROSOL EFFECTS

All aerosols reflect the sun's radiation like tiny mirrors, increasing the earth's albedo and providing a direct cooling effect to the earth's surface. The magnitude of this cooling effect will depend on the aerosol's number density, size distribution, altitude, and lifetime in the atmosphere.

There are additional interactions with radiant energy from both the sun and the earth. The nature of this interaction depends on the chemical composition of the aerosols and may lead to either a cooling or a warming effect. If the chemicals in the aerosol absorb the radiation that comes from either the sun or the earth, the effect can be atmospheric warming, but if the chemicals only reflect the radiation, the effect will be atmospheric cooling.

The very small particles can also act as cloud condensation nuclei, which are specks of microscopic dust on which water vapor can condense and contribute to cloud formation. As we discussed in Chapter 8, clouds generally have a cooling effect, but they can also cause warming under certain circumstances. Aerosols will initiate cloud formation when the humidity is right, the aerosols are in the right size range, and they reach the right altitude. This cloud formation will occur regardless of the source of the particles.

Remember that over 60% of the world is covered by clouds at any one time. In his 1994 *Nature* article, "Dirty Clouds and Global Cooling," Graeme Stephens, of the Department of Atmospheric Sciences at Colorado State University, summarized the growing evidence that clouds that are "dirty" from man-made aerosols provide a greater cooling effect than other clouds.[5] The evidence indicates that there are more, but smaller, particles in dirty clouds, which are more efficient at reflecting the sun's radiation. Therefore, these particles increase the albedo of the polluted clouds. Stephens states: "When all effects of anthropogenically produced aerosol are taken together, then the predicted global forcing is comparable in magnitude to the direct forcing by greenhouse gases."[6] This means simply that computer climate models can predict a cooling effect of greenhouse gases that is equal to the predicted warming effect! Stephens goes on to point out that the effects of aerosols are primarily regional and the global effects are subtle. All these cloud-related factors complicate the overall aerosol effect on the global climate. The dynamics and subtle effects of aerosols on the fluctuating cloud cover of the globe have been inadequately dealt with in computer climate models.

SULFURIC ACID AEROSOLS

It is important to distinguish between the aerosols in the stratosphere and those in the troposphere. Aerosols in the stratosphere, the layer of atmosphere above the troposphere, arise from different sources. They must be treated differently by climate modelers from those in the troposphere because of their distinct chemical composition, circulation dynamics, radiational interactions, and many other factors.

The primary aerosol affecting the climate in the stratosphere is a sulfuric acid aerosol that consists of about 75% sulfuric acid (H_2SO_4) and 25% water, according to James Rosen of the University of Wyoming and Vladimir Ivanov of the Main Geophysical Observatory St. Petersburg, Russia, in their excellent chapter "Stratospheric Aerosols."[7] Nearly all sulfur-bearing aerosols become sulfuric acid if they reach the stratosphere. These aerosols come from both natural and anthropogenic sources, but most that reach the stratosphere come from volcanoes, according to D. J. Hofmann.[8] Hofmann spent over twenty years studying atmospheric aerosols by means of balloon-borne instrumentation at the University of Wyoming before joining NOAA's Climate Monitoring and Diagnostics Laboratory at Boulder, Colorado.

Aerosols in the troposphere usually consist of a mix of the various particle types, such as sulfuric acid, soot particles, or insect parts. The tropospheric aerosols have a significant anthropogenic component, most of which comes from sulfur dioxide emissions associated with combustion of fossil fuels. See Table 9.2 for an estimate of sulfur-containing aerosols from different sources. Sulfur dioxide gas interacts with solar radiation and water vapor to form sulfuric acid aerosol.

Volcanoes

For many years the occurrence of a cold summer following a large volcanic eruption has been attributed to the emission of volcanic aerosols. Perhaps the first person to make this observation was Benjamin Franklin, following the great Laki eruption in Iceland and the Asama eruption in Japan in 1783. There are historical references to years with no summer and to years with summer snow in areas with typically very warm summer weather. But it has

Table 9.2. Summary of Sulfur Containing Emission Sources[a]

Source	Chemical form	Tg[b] per year
Volcanic eruptions	mainly SO_2[c]	7–10
Oceans	DMS[d]	10–50
Soils and plants	DMS & H_2S[e]	0.2–4
Biomass burning	SO_2	0.8–2.5
Anthropogenic emissions	SO_2	70–80

[a]J. T. Houghton, B. A. Callander, and S. K. Varney, eds. *Climate Change 1992* (Cambridge: Cambridge University Press, 1992) p. 41.

[b]A Tg is a teragram or 1×10^{15} grams (One Tg = One GT, gigaton, a billion tons).

[c]SO_2, sulfur dioxide gas, the major source of sulfur aerosols in the atmosphere.

[d]DMS stands for dimethylsulfide, which is a gaseous chemical given off by plankton and certain plants and organisms in the soil.

[e]H_2S, hydrogen sulfide gas, is another sulfur-containing gas given off by certain organisms in the soil.

been only in modern times that scientific studies of volcanic emissions have provided the beginnings of a real understanding of the atmospheric effects of volcanic activity.

The chemical composition of the gases and particulates emitted by volcanic eruptions varies with the geology of the location. The major gas is sulfur dioxide (SO_2), but hydrogen chloride (HCl) and hydrogen fluoride (HF) gases are also emitted in large quantities, and there are many other chemicals emitted in smaller quantities. Rosen and Ivanov estimate that volcanoes produce from two to ten million tons per year of SO_2, about two million tons per year of HCl, and about 100,000 tons per year of HF.[9] These gases are ejected explosively by the volcanic eruption, which means that volcanoes eject a great deal of sulfur dioxide gas directly into the stratosphere in addition to the dust and gas that remain in the troposphere. In the stratosphere the SO_2 is chemically changed into a sulfuric acid aerosol that can reflect sunlight back into space and thus directly cool the earth.

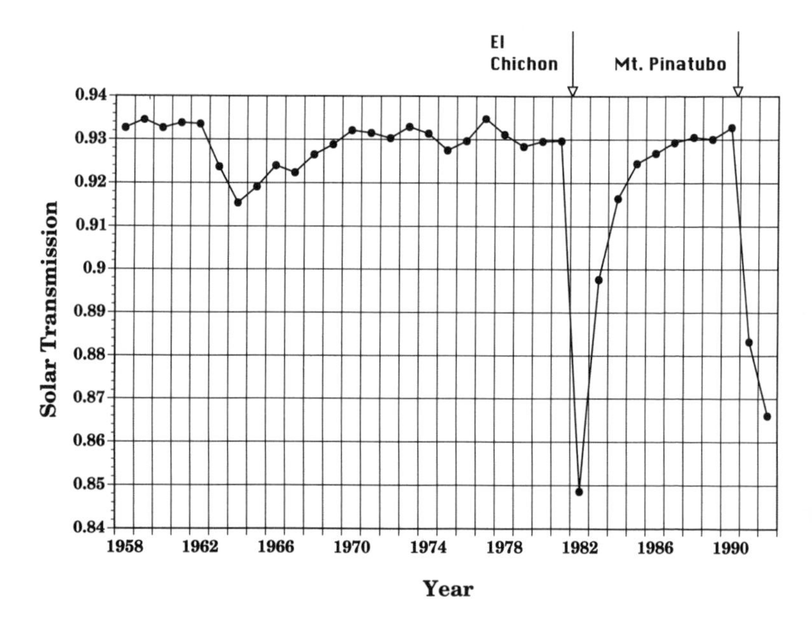

FIGURE 9.1. Sunlight transmitted through the atmosphere from 1958 to the present as measured at Mauna Loa, Hawaii. Each point represents the average solar transmission for the year. A transmission value of one represents 100%. This figure was adapted from data in E. G. Dutton, "Atmospheric Solar Transmission at Mauna Loa," in *Trends '93: A Compendium of Data on Global Change* ORNL/CDIAC-65, ed. T. A. Boden, D. P. Kaiser, R. J. Sepanski, and F. W. Stoss (Oak Ridge, TN: Carbon Dioxide Information Analysis Center, Oak Ridge National Laboratory, 1994), pp. 479–482.

This direct cooling effect is illustrated in Figure 9.1, which shows the amount of sunlight transmitted to the earth from 1958 to the present, as measured at Mauna Loa, Hawaii, by E. G. Dutton and others from NOAA's Environmental Research Laboratories.[10] Note the dramatic decreases in solar transmission after the El Chichón event in 1982 in Mexico and after the famous Mt. Pinatubo event in the Philippines in 1991, along with the less dramatic effect of the Agung eruption in Bali, Indonesia in 1963. It is important to observe that the effects of these large volcanoes last for several years. The effect of El Chichón lasted at least four years, and the aerosol particle count was not yet back to original levels when Mt. Pinatubo erupted in 1991.

In the past, scientists thought that volcanic aerosols had short lifetimes, maybe a few weeks, and that therefore they had no major effect on the long-term global climate. However, during the past twenty or so years, measurements were made of the amount and size of aerosol particles in the stratosphere, using balloon-borne particle counters sent aloft from a facility at the University of Wyoming.[11] These measurements indicate the buildup and decay of volcanic aerosols in the stratosphere. The graph of these data shows the number of particles plotted as a function of time; it resembles Figure 9.1. The increase of aerosol particles in the stratosphere when the volcanic event occurs is impressive, but more importantly, the slow decay of the numbers of these particles indicates a much longer effect than scientists had previously thought. The University of Wyoming is directly in the path of the Mt. St. Helens plume, which accounts for the large increase in particulates measured after its eruption.

Mt. Pinatubo was one of the three largest volcanic events of the twentieth century.[12] It ejected about 3 cubic kilometers of material, which is about six times that of El Chichón, and it spewed forth about three times more sulfur dioxide (SO_2). The SO_2 cloud circled the earth in about three weeks and spread into both hemispheres. It was such an impressive event that *Geophysical Research Letters*, an important research journal for climate scientists, devoted a special section in 1992 to the early findings about the eruption.

Papers in this special section included a very interesting contribution from Jim Hansen and a group of researchers from NASA Goddard Space Flight Center. (Of course you remember that it was Hansen who testified to Congress that he was "99% sure" that global warming was upon us in 1988.) Hansen performed global temperature experiments using his computer climate model with an input of SO_2 of twice that emitted by El Chichón. The results of his model predicted a decrease in temperature (cooling) of about 0.5°C (0.9°F), occurring in the 1991–1993 period.[13] Measurements of SO_2 from Mt. Pinatubo suggest that there could be three times the emissions as compared to El Chichón. If this is true, Hansen's computer climate models would have indicated additional cooling.

Hansen was careful to point out the many uncertainties and possible errors involved in the computer models in his paper, as most scientists do in their technical publications. For example, Hansen states that "a key mechanism which could limit the response to a negative climate forcing [cooling], heat exchange with the deep ocean, is simulated very crudely in our model." He goes on to say that he needs more information about the nature of the sulfuric acid aerosol size distribution and the effects of the aerosol on the ozone layer. He is not sure if his model can handle the injection of water vapor into the stratosphere, and "climate feedbacks, such as changes of cloud properties and atmospheric water vapor, may not be accurately simulated in our climate model."[14] Nonetheless, Hansen's prediction of 0.5°C cooling from the Mt. Pinatubo eruption is widely quoted in the scientific literature, including the 1992 IPCC report.[15]

Hansen states that the effects of "Pinatubo aerosols should provide an acid test of climate models."[16] This means that he feels the climatologists should be able to detect these effects in the climate and thus verify the computer climate models. Only time will tell.

There is a natural level of sulfuric acid aerosol in the stratosphere due to the ongoing volcanic activity on earth. According to Hofmann, there is evidence that this natural background in the stratosphere is increasing.[17] If this is the case, climate scientists have yet another perturbation, or factor, that is unaccounted for in their models. Figure 9.1 shows the aerosol effects from really large volcanoes. The fact that the sun's radiation is decreased substantially after a large volcanic eruption is direct experimental evidence of the cooling effect.

Our knowledge of volcanic activity is incomplete; we can predict neither when nor how large a volcanic eruption will be. However, it is well known that the earth is geologically quite active, with its large, continental plates of earth mass continually pressing against each other, trying to relieve the planet's internal stresses. The actions of plate tectonics cause many smaller active volcanoes as well as the occasional major one. During the ten years between 1974 and 1984, for example, there were over seventy volcanic eruptions classified as "large."[18] An earth that is producing over seven large volcanic eruptions per year may have a condition of naturally increasing stratospheric aerosol background, a condition that should be included in any computer global climate model.

As stated above, the aerosol component of the stratosphere is believed to be essentially from natural sources (mainly volcanic eruptions), with only a hint from anthropogenic sources. On the other hand, it would appear that much of the aerosol in the troposphere may be anthropogenic. Data in Table 9.2 provide estimates of sulfur emissions from all sources, most of which do not reach the stratosphere. Sulfur aerosols that do reach the stratosphere have a much larger direct climate effect than those contained in the troposphere.

Ocean-Derived Aerosols

Aerosols are produced at the ocean surface in a variety of ways. White caps caused by windy weather consist of millions of bubbles that burst into tiny droplets, many of which are so small that they are carried into the atmosphere. An aerosol of sea salt consisting mostly of sodium chloride (the same chemical as table salt) results. These salt particles are generally coarse and short-lived and do not affect long-term weather patterns. However, if global temperature does rise and the ocean winds increase, as predicted by the computer climate model of J. Latham and M. H. Smith of the University of Manchester in England, the increase in wind velocity will decrease the size of the sea salt aerosols. This decrease in size will create additional cloud-forming nuclei and increase the ocean clouds and the resultant cloud albedo (radiation reflection) to the point of offsetting global warming.[19] These predictions have been derived from experimental measurements of the size distributions of sea-salt aerosol as related to the wind velocity. Whether this prediction of increased winds will prove accurate must wait on the "predicted" global warming; however, it presents another factor that computer models will need to include to be valid. Latham and Smith did not include the effects of the sea-salt aerosol on the dimethyl sulfide that occurs naturally from the oceans (Table 9.2) in their models.

Phytoplankton, the fundamental core of the ocean's food chain, generate a sulfur-bearing gas, dimethyl sulfide (DMS), as a product of their respiratory activity. The extent of DMS and its effects on cloud formation and the climate was the focus of a review article in *Nature* by a group of scientists, including Robert Charlson at the University of Washington and James Lovelock of Coombe Mill Experimental Station at Cornwall, England.[20] They relate that DMS gas is produced by a number of different phytoplankton types. In addition, algae secrete sulfur compounds that react in the ocean to produce even more DMS. The DMS is short-lived in the atmosphere, with the final chemical product being sulfuric acid. Thus, sulfuric acid aerosols are indirectly generated in the oceans, and they account for most of the naturally occurring sulfuric acid aerosol in the troposphere.

Charlson's paper concludes that there is little indication that the sulfuric acid from ocean aerosols ever reaches the stratosphere, but there is evidence that this source of sulfuric acid aerosol participates in cloud formation and contributes to increased cloud albedo in the warmer ocean regions near the equator.[21] This, of course, means that the ocean-generated aerosol may have a larger cooling role than has been thought and represents yet another factor not included in most computer climate models.

Anthropogenic Sulfur Emissions

Atmospheric sulfur emissions directly related to human activities make up well over half of all sulfur emissions (Table 9.2). Most of these emissions are related to burning sulfur-bearing fuels or to processing sulfur-bearing ores or minerals to obtain base metals such as iron and copper. They are directly related to the industrialization of countries, and, according to Nicholas Lenssen, on the staff of the Worldwatch Institute, the developing countries are increasing these emissions at a much greater rate than the industrialized countries.[22] These emissions, along with carbon dioxide and other greenhouse gases that are fundamental to the quality of life, are related to the size of the world's population.

It is not clear whether the global cooling effects due to the resulting sulfuric acid aerosols will offset the warming effects due to greenhouse gases. There will be believable arguments on both sides of the issue. However, it is clear that humans are responsible for emitting gases that cool the earth along with gases that warm it.

REGIONAL SOURCES AND EFFECTS

There is a much larger concentration of particulate matter in the atmosphere over the continents than over the oceans or at the poles (Table 9.1). This is especially true of very small particles, the size that can act as cloud condensation nuclei. There are from 10 to 1,000 times as many small particles per cubic centimeter over the landmasses as over the oceans. About 1,000 of these particles are due to dust in remote barren areas, as indicated by the "remote continental" heading in Table 9.1. There is an increase to nearly 9,000 particles per cubic centimeter in "rural" areas, presumably due to farming activities. The major source is clearly associated with the large, dense populations of the world's cities, however, which produce well over 130,000 particles per cubic centimeter.

In the remote and rural areas, particulate matter consists mostly of very small silica particles of common sand, along with decaying insects and other bits of biological matter. In rural agricultural areas the soil is exposed to winds between crops. The necessity of plowing the fields before planting crops increases the soil-containing aerosol over these regions. Furthermore, it is common practice in many parts of the world to burn off vegetative growth between plantings. Burning of any biomass creates carbon soot particles and sulfur dioxide gas, in addition to carbon dioxide and water vapor.

Enter the cities and you find automobiles, incinerators, electric power generation plants, factories, refineries, and other chemical processing plants. These sources emit various kinds of gases and particulates. The most prominent particulates are sulfate particles originating from SO_2 gas emitted by fossil fuel combustion. Some of these are large and last only a few hours or days in the atmosphere before arriving on someone's dresser or bookshelf as "ordinary dust." But many of the smaller particles will float with the winds and stay in the atmosphere for weeks. Regardless of size, during its residence time in the atmosphere, each particle acts as a mirror that directly reflects the sun's radiation.

Two important considerations must be taken into account with respect to these aerosol sources. The first is that the major aerosol concentration is associated with the urban centers of the world. The second is that nearly all urban centers are in the Northern Hemisphere, so the primary cooling effect from these anthropogenic aerosols will be more prevalent in the Northern Hemisphere. More complication is thus added to computer global warming models.

RECENT STUDIES

A 1992 *Science* article written by a group of climate scientists led by Robert Charlson from the University of Washington, with scientists from several other organizations, including Jim Hansen, estimated that tropospheric aerosol cooling equals the warming caused by CO_2 emissions.[23] It is important to understand that this paper addresses cooling due to aerosols in the troposphere, whereas the paper that Hansen wrote about sulfur dioxide emissions from Mt. Pinatubo was concerned with the cooling effects of the aerosol in the stratosphere. These are two different cooling effects.

In recent years articles in the scientific literature have described at least four climatic effects from aerosols:

- Stratospheric cooling caused by volcanic injection of SO_2.

- Direct tropospheric cooling caused in part by man's burning of fossil fuels and industrial processing of metals.
- Enhanced cooling caused by "dirty" clouds.
- Potentially enhanced ocean aerosol generated by the increased winds associated with global warming.

Each of these cooling effects is purported to be equal in magnitude to the global warming caused by greenhouse gas emissions.

If we add up all these cooling effects, we should be concerned—but not about global warming. All these effects have not yet been included in computer climate models. It will be interesting to see the results of predictions when everything is included. Which will it be—global warming or global cooling?

REFERENCES

1. J. T. Houghton, B. A. Callander, and S. K. Varney, eds., *Climate Change 1992* (Cambridge: Cambridge University Press, 1992), pp. 40–42, 62–64.
2. R. J. Charlson, J. Langer, H. Rodhe, C. B. Leovy, and S. G. Warren, "Perturbation of the Northern Hemisphere Radiative Balance by Backscattering from Anthropogenic Sulfate Aerosols," *Tellus* 43AB (1991):152–163.
3. S. I. Rasool and S. H. Schneider, "Atmospheric Carbon Dioxide and Aerosols: Effects of Large Increases on Global Climate," *Science* 173 (1971):138–141.
4. G. M. Krekov, "Models of Atmospheric Aerosols," in *Aerosol Effects on Climate*, ed. S. Gerard Jennings (Tucson, AZ: University of Arizona Press, 1993), pp. 12–13.
5. Graeme L. Stephens, "Dirty Clouds and Global Cooling," *Nature* 370 (1994):420–421.
6. Ibid.
7. James Rosen and Vladimir A. Ivanov, "Stratospheric Aerosols," in *Aerosol Effects on Climate*, ed. S. Gerard Jennings (Tucson, AZ: University of Arizona Press, 1993), p. 168.
8. D. J. Hofmann, "Increase in the Stratospheric Background Sulfuric Acid Aerosol Mass in the Past 10 Years," *Science* 248 (1990):996.
9. Rosen and Ivanov, 157–158.
10. E. G. Dutton, "Atmospheric Solar Transmission at Mauna Loa," in *Trends '93: A Compendium of Data on Global Change*, ORNL/CDIAC-65, ed. T. A. Boden, D. P. Kaiser, R. J. Sepanski, and F. W. Stoss (Oak Ridge, TN: Carbon Dioxide Information Analysis Center, Oak Ridge National Laboratory, 1994), pp. 479–482.

11. James Rosen and Vladimir A. Ivanov, "Stratospheric Aerosols," in *Aerosol Effects on Climate*, ed. S. Gerard Jennings (Tuscon, AZ: University of Arizona Press, 1993), p. 174.
12. Peter Francis, *Volcanoes, A Planetary Perspective* (Oxford: Clarendon Press, 1993), p. 384.
13. James Hansen, Andrew Lacis, Reto Ruedly, and Makiko Sato, "Potential Climate Impact of Mount Pinatubo Eruption," *Geophys. Res. Lett.* 19 (1992):215–218.
14. Ibid.
15. Houghton, Callander, and Varney, 62–64.
16. Hansen, Lacis, Ruedly, and Sato, 215–218.
17. Hofmann, 996.
18. Elmer Robinson and Thomas E. Defoor, "Stratospheric Aerosol Conditions over Mauna Loa during Recent Quiescent Volcanic Periods," in *Aerosols and Climate*, ed. Peter V. Hobbs and M. Patrick McCormick (Hampton, VI: A. Deepak Publishing, 1988), pp. 328–329.
19. J. Latham and M. H. Smith, "Effect on Global Warming of Wind-Dependent Aerosol Generation at the Ocean Surface," *Nature* 347 (1990):372–373.
20. Robert J. Charlson, James E. Lovelock, Meinrat O. Andreae, and Stephen G. Warren, "Oceanic Phytoplankton, Atmospheric Sulphur, Cloud Albedo and Climate," *Nature* (1987):655–661.
21. Ibid.
22. Nicholas Lenssen, "Providing Energy in Developing Countries," *State of the World 1993* (New York: W.W. Norton, 1993), pp. 103–104.
23. R. J. Charlson, S. E. Schwartz, J. M. Hales, R. D. Cess, J. A. Coakley, Jr., J. E. Hansen, and D. J. Hofmann, "Climate Forcing by Anthropogenic Aerosols," *Science* 423 (1992):423–430.

PART IV

Prediction or Reality?

TEN

The Truth Behind the Myth

If Science is to progress, what we need is the ability to experiment, honesty in reporting results—the results must be reported without somebody saying what they would like the results to have been—and finally . . . the intelligence to interpret the results.

—RICHARD FEYNMAN,
NOBEL PRIZE IN PHYSICS, 1965
THE CHARACTER OF PHYSICAL LAW[1]

Dr. Feynman was one of the outstanding scientists of this century, whose death in 1988 ended his distinguished career as a theoretical physicist, much of it spent at the California Institute of Technology. Feynman believed that scientists must experiment and judge the results of their experiments in an objective and unbiased manner. As honest, open-minded researchers, they cannot simply perform experiments to "prove" their preconceived notions.

After the terrible accident of the *Challenger* space shuttle in 1986, Dr. Feynman was part of the investigative team that was formed to determine the cause. He thought that the evidence pointed to failure of an O-ring that was used to seal two parts of the body of the rocket to prevent fuel leakage. But many engineers and scientists contended that the O-ring material, designed to operate properly under the wet, freezing conditions that were encountered at takeoff, could not have failed under these circumstances. In fact, in spite of acknowledgment of the criticality of the sealing problem, a NASA evaluation stated: "Analysis of existing data indicates that it is safe to continue flying existing design [of the *Challenger*] as long as all joints are leak checked . . ." Feynman found out what the "analysis of existing data" was. He states:

> It was some kind of computer model with various assumptions that were not necessarily right. You know the danger of computers, it's called GIGO: garbage in, garbage out? The analysis concluded that a little unpredictable leakage here and there could be tolerated, even though it wasn't part of the original design.[2]

Yes, the disaster of the *Challenger* was in part caused by the inadequacy of a computer model and a misplaced belief in its prediction!

Dr. Feynman performed a simple experiment for the investigative committee with an O-ring made of the suspect material. Placing the O-ring in a glass of ice water, he demonstrated that it would fail under wet, icy conditions. He performed the seminal experiment!

In the final analysis, computer models must pass this test of empirical evidence. To be considered valid, computer models must correctly predict results that can be verified by actual experiment or observation. Computer climate models have so far failed to do this.

What have we learned from this discussion about global warming? Thanks to the more than five billion people in the world today and their needs for food, clothing, warmth, and an improved quality of life, the atmosphere has been altered by a small, but significant, amount. Certain gases (such as carbon dioxide and methane), which are absolutely necessary for human existence, have increased in concentration in the atmosphere. These gases interact with radiative energy from both the sun and the earth to create the greenhouse effect that helps warm the earth.

The greenhouse effect is essential to our survival; without it the earth would be too cold for human habitation. However, some scientists and many environmental activists warn that increases in anthropogenic greenhouse gases might cause a runaway global warming effect. But can we be certain that carbon dioxide levels in the atmosphere are actually the driving force for climate change? It has been pointed out that a decrease in levels of CO_2 during past ice ages has lagged behind the decrease in global temperatures. Does the temperature of the earth determine the carbon dioxide level or does the carbon dioxide level determine the temperature of the earth?

Nevertheless, all computer climate models that have been developed by climatologists predict some degree of warming, given the scenario of doubling the levels of greenhouse gases in the atmosphere. There are many different computer climate models, and they each give a different prediction of increased warming. Some of them predict a large warming, up to 5.3°C (9.5°F), others a small warming, about 1.7°C (3.1°F), and still others predict temperature increases between these extremes. Some of the models predict that a doubling of greenhouse gases will happen by 2050, others by 2090.

Model predictions of global temperature increase are moderated when the level of carbon dioxide is increased gradually, rather than abruptly. An abrupt doubling of carbon dioxide is totally unrealistic! It has been calculated that 50 to 100 years of ever-increasing burning of fossil fuels will be required to double the level of carbon dioxide in the atmosphere.

Further, none of the current computer climate models can predict much about future regional weather because they are too crude in their spatial resolution (each area segment is too large), and yet such regional information is critical to policymakers.

Nonetheless, because of public pronouncements by some scientists, copious propaganda provided by environmental organizations, and zealous reporting by the media, a popular myth has evolved about global warming, prompting institutional calls for action. This myth is partly based in fact, but it is exaggerated in detail. The exaggerations have been propagated by some scientists, perhaps because of the need for popular and political support to fund the very expensive climate research.

In the past, the scientific community usually counseled for calm and moderate action when the end of the earth was predicted due to some perceived impending disaster; however, in the case of global warming, it appears that many scientists have been swayed by the calculations of their computer models to announce that disaster is just around the corner. This pronouncement is being made in spite of the many known limitations and shortcomings of computer climate models.

One of the purposes of this book is to bring to your attention these limitations and shortcomings and to caution you not to overreact to the predictions of doom. Many of the remedies being proposed to prevent global warming could in fact create greater global problems if carried out. These remedies could also cause economic hardship to both the industrialized nations and the developing ones. For example, planting trees is a good idea for many reasons; however, devoting an area one-fifth the size of the United States to intensive forestry is a little extreme! The idea of spending billions of dollars and tying up most of the commercial aircraft in the world to spread particulates in the stratosphere is not only incredible, it could cause serious global cooling!

In the following sections we would like to summarize the significant arguments and facts concerning the uncertainties associated with the global warming myth, the detection of global warming, and the unscientific extrapolation of computer models beyond their realistic capability to provide useful predictions.

DETECTING GLOBAL WARMING

Unfortunately, there was no way to measure the earth's average temperature accurately until the development of satellite-based weather-monitoring probes. Satellite measurements started in the late 1970s and global temperature has been monitored since then. During this period, short-term fluctuations have been detected, but no general global warming trend has been established. If the predictions of computer climate models were correct, there should have been an increase of over half a degree during this measurement period (0.5°C, or 0.9°F).

Intense efforts were made before the satellite age to develop a procedure to determine the earth's temperature by making use of weather data from the many weather stations around the world. This process undoubtedly has provided reasonably good results for the past few decades, but it falls woefully short for previous periods, for these reasons:

- There were too few stations.
- The stations were not spread evenly over the globe and large portions of the globe were not adequately measured.
- Methods and procedures used for calibrating, recording, and measuring changed over the years.

If the observational area is restricted to a region that was adequately measured with weather stations during the past century, such as the United States, no warming trend is observed (see Figure 5.6).

This brings us to a conclusion that is not in agreement with the assessment of the International Panel on Climate Change (IPCC), and that is that the earth has not warmed by 0.3–0.6°C (0.5–1.1°F) during the past 100 years. Or, if the earth has warmed, the data are insufficient to detect this warming in the presence of the very large temperature variations from day to night, summer to winter, and decade to decade.

The temperature variation from decade to decade is much smaller than seasonal changes, and yet, it was only a few years ago, in the 1970s, that a decade-length fluctuation caused many scientists to think a global cooling had begun.

LIMITATIONS OF COMPUTER CLIMATE MODELS

We have learned that computer climate models are an incredibly complex set of mathematical equations that attempt to describe all aspects of the global climate. These models divide the earth's atmosphere into calculation volumes, or cells, that are then integrated to predict what the weather will be in the cell at some time in the future. This entails thousands of cells and enormous numbers of computations and stresses the limits of modern-day supercomputers. It can take weeks of continuous computer run-time for a 100-year global climate simulation. This is very expensive and scientists are forced to take shortcuts to save computer time and money. The expense explains why the calculation cell dimensions have been so large, with an area about the size of New Mexico (about 310 by 400 miles, or a 124,000 square mile area) being the smallest observable segment.

Recently, the U.S. National Weather Service (NWS) expanded its official forecasts to include a fifteen-month view. Edward O'Lenic, of the NWS operation at Camp Springs, Maryland, states, "We're not forecasting the weather; we're forecasting the climate." Remember that the definition of climate is the average weather for a period of time. This means that the NWS is in the climate-modeling business. The NWS uses a computer climate model very similar to the ones on which global warming predictions are based, except that it operates on a more regional spatial level. O'Lenic goes on to comment on the reliability of the predictions: "If you cannot accept forecasts [that are right only] five-and-a-half to seven times out of ten, then you shouldn't be using these forecasts."[3] The NWS scientists admit that they are stretching the capability of their climate model for this forecast. Take note that the NWS predictions are for only fifteen months, not the 50–100 years that other climate modelers are projecting for their global warming predictions.

Further, it has been emphasized time and again that many features that must be included in computer models exhibit chaotic behavior. One of the main properties of a chaotic system is its unpredictability; another is that small changes in the input data generate large differences in the calculated results. This property makes extrapolation into the future extremely risky. These chaotic features appear in ocean currents, cloud formation and dynamics, general air circulation in the atmosphere, and in many other important variables that must be addressed by the models.

There are serious gaps in the ability of scientists to determine the mathematical relationships that accurately describe particular critical features of the climate. This failure in the mathematical description results in an inability to predict accurately the effects of these features. We have discussed the complex interaction of water vapor and particulate aerosols in cloud formation, the even more complex interaction of all the different cloud types with radiation generated by the sun and the earth, the dynamics of cloud formation and dissipation, and the interface between the oceans and atmospheric circulations. None of these major features is handled adequately by computer models. When you add to this the general lack of predictability of important natural climatic events such as volcanic eruptions, hurricanes, typhoons, and El Niño events, you must be skeptical of the value of the results of fifteen-month projections, and even more so of 50- or 100-year projections.

Remember, too, that the modelers are adjusting the levels of carbon dioxide somewhat arbitrarily. Although data are available on the quantities of anthropogenic emissions of carbon dioxide and other greenhouse gases into the atmosphere, measurements of these gases indicate that the actual levels are lower than they should be! There should be more carbon dioxide in the atmosphere than scientific instruments indicate. What has happened to the missing carbon? The scientists do not know where all of these gases have gone! Their understanding of the carbon sinks in our oceans, forests, and soils is incomplete.

Finally, computer modelers have set some features of their models at a constant value even though these features vary, or fluctuate, with time. One of these features is solar energy, which is known to fluctuate. In the few experiments that have included this variation, the predicted global warming has not been as great as when it was left out. Many models set the albedo, or solar reflection, of the clouds as a constant, even though it is not. Many modelers input data only for summer or winter and do not attempt to model the full annual cycle. The scientists who publish results from these inadequate models generally list the details of their assumptions and simplifications in their articles, but they all too often have an unrealistic faith in the results of computer calculations in spite of this knowledge. This is human nature, but it lacks scientific objectivity. The saying common among computer modelers, GIGO ("garbage in, garbage out") is very appropriate.

When the results of technical reports are presented to the public by the media, the cautions and assumptions stated in these technical papers are omitted. The media thus contribute to the popular myth that terrible and mighty change is about to affect the world climate.

We are not convinced that the computer models are advanced to the stage that their projections of 50–100 years have sufficient validity to sway an objective scientist to push for governmental action of great consequence, such as making drastic cuts in fossil fuel consumption, or initiating other economically disastrous action. The fact is the earth should already have warmed enough for detection by our current experimental measurement systems if the computer climate models were correct. This warming has not occurred, which is reasonable proof of their inadequacies.

Just as the early predictions of great 25-foot sea level increases have gradually become smaller and smaller with more refined (and better) input data, the predictions of global warming are also decreasing with more realistic input data. Although no modeler has yet put all of the moderating factors into one simulation experiment, each experiment that has included a more realistic approach has predicted a smaller global warming.

The argument that "we must take drastic action because it *might* be true" is nonsense. Dr. Feynman's caution that scientists should not be sure of the outcome of their results ahead of time, or before they examine the experimental data, is an important tenet that has unfortunately been ignored by many in the scientific community.

WHAT IS REALLY HAPPENING: GLOBAL WARMING OR GLOBAL COOLING?

Since climatologists and environmental scientists cannot detect evidence of global warming, they now seem to be finding all sorts of cooling influences to explain the lack of warming. Some have blamed the weather (El Niño) for the lack of important evidence of warming. It is clear that scientists overlooked the importance of natural events such as volcanic eruptions in their computer models. Mt. Pinatubo did indeed create a stratospheric aerosol that had global impact, mostly a cooling effect. There is evidence that stratospheric aerosol background is increasing. There is also evidence that tropospheric aerosols are increasing due to anthropogenic sulfur dioxide emissions—another cooling effect. There is even recent evidence that the thinning ozone layer (due to anthropogenic CFCs in the stratosphere) is causing cooling in the troposphere.

If all of these reasons for global cooling are as important as suggested, there should be concern that the cooling effects will overwhelm the warming effects. Maybe government policy should be to increase carbon dioxide emissions to offset the dire consequences of global cooling (just joking, of course!).

IS GLOBAL WARMING REALLY BAD?

Not all of the scientific reports about increased levels of carbon dioxide in the atmosphere are apocalyptic. There are scientists in the agricultural research community who argue that increases in carbon dioxide coupled with a slightly warmer climate will be good for plant life. Plants are the fundamental core of the food chain. One might suggest that what is good for vegetation is good for humankind! The importance of carbon dioxide to plants is not based on the projections of computer models, but on empirical studies with live plants. Increases of plant yield with increased carbon dioxide levels have been measured with a great many types of plants by a large number of agricultural researchers. Remember that carbon dioxide is essential for photosynthesis, the process by which plants grow. Without carbon dioxide there would be no plants, and animals, including humans, could not exist.

POLICY CONSEQUENCES

Political action is being increasingly influenced by science. Unfortunately, in some cases scientific results may be swayed by political action. It certainly can be proven that research funding in areas that are in political favor is more abundant than in areas that are out of political favor. We have discussed the admitted hope of environmental groups and scientists that the weather will be particularly warm to help get the U.S. Congress excited about global warming. The environmental activists hope to incite Congress into environmental legal action, and the scientists hope to incite Congress into increased funding for environmental research.

Regardless of the motivation, there is real public and political concern about the possibility of runaway global warming, and the United States has signed an international agreement to reduce carbon dioxide emissions to 1990 levels by the year 2000. This treaty, discussed in Chapter 3, was the result of the Rio Earth Summit. Its official title is The United Nations Framework Convention for Climate Change, and it was signed by 160 countries. The ultimate goal of the treaty is to stabilize carbon dioxide and other greenhouse gas emissions at a level that "would prevent dangerous anthropogenic interference with the climate system." According to the treaty, the reduction is to apply only to the industrialized countries, with the developing countries to start compliance at a later date.

Pamela Zurer, a science reporter for *Chemical & Engineering News*, reported on a December 1994 climate conference in Washington, D.C., sponsored by the Center for Environmental Information. In this report she quotes Daniel Reifsnyder, director of the Office of Global Change at the State Department, as follows, "It's no secret that meeting this goal [of the treaty] is very difficult. . . . Some nations had given very little thought as to how specifically to meet the goal."[4] Of course, the energy needs of the developing countries will increase as their populations increase and if their quality of life increases. So the anthropogenic emissions of carbon dioxide and other greenhouse gases will continue to increase, in spite of any measures taken by the industrialized nations.

Some countries already tax the use of fossil fuels in order to decrease carbon dioxide emissions. In fact, some of these taxes often constitute a large percentage of the cost of the fuel. According to Zurer's report, Germany's gasoline tax is two-thirds of the cost of gasoline. She quotes Wolfram Schoett of the German Embassy in Washington, D.C., as stating, "I was shocked when I went home in October to find it cost $40 to fill up my son's VW Golf."[5] Germany's stated goal is a 50% reduction in 1987 levels by 2005. In the United States, protests of these fuel-related taxation policies is increasing. When put on the ballot, many proposed increases are voted down. People seem to be willing to endorse reduction of greenhouse gases until it comes to their pocketbooks.

It is highly unlikely that the goal of reducing carbon dioxide emissions to 1990 levels will occur in the year 2000, or anytime, without drastic economic impact. Is that impact warranted? The current state of scientific knowledge cannot provide a solid foundation for the answer. At this point, after an open-minded evaluation of the scientific facts, it cannot honestly be stated that global warming is happening. We cannot say that the earth is any warmer today than it was 100 years ago, nor can we say that it would be bad for humanity if it were!

WHAT CAN WE DO SOMETHING ABOUT?

Maybe our government and the environmentalists should focus on issues that are more manageable and more important to our future well-being, such as population control, pollution abatement, waste treatment, conservation of natural resources, development of cleaner energy sources, and development of more energy efficient power sources. In many ways the increases in greenhouse gases, air pollutants, and wastes, and the depletion of our natural, nonrenewable resources are intimately connected to the ever-increasing population.

Scientists will continue to focus their research in areas that receive government funding. Clearly, improvement in the ability to model our global climate is important, but the areas mentioned above are also important and deserve governmental funding and focus. The government should be extra cautious in dictating the specifics of research, however. It is one thing to identify an area of focus, but it is entirely different to specify the details. Many major scientific breakthroughs are not deliberate; a scientist often "stumbles over" an important discovery in one field while doing research in another.

This type of accidental discovery is termed *serendipity,* and has been quite common throughout the ages. Sir Isaac Newton's realization of the law of gravity in the mid-1600s was reportedly triggered by his observation of an apple falling from a tree. Wilhelm Roentgen accidentally observed X-rays from the cathode-ray tube in 1895. In 1886 Henri Becquerel accidentally exposed a photographic plate to a radioactive substance and discovered radioactivity. In 1865 Friedrich Kekule uncovered the secret of the cyclic structure of benzene after having a dream of a snake biting its tail. Numerous stories of accidental scientific discoveries of all kinds are related in Professor Royston Roberts's very entertaining book *Serendipity, Accidental Discoveries in Science.*[6] Scientists must be given latitude to follow their curiosity.

The moral of this book, if there is a moral, is that informed citizens, students, and politicians should be concerned about our environment. They should support conservation programs that preserve our natural resources, protect our atmosphere from pollution, help to eliminate wastes, and recycle potential wastes. But before hysteria sets in, they should learn the facts and not be swayed by the exaggeration and fear tactics of special-interest organizations whose motives are self-serving and do not have the interest of the general public in mind. It is critical to understand both sides of any issue and to act with intelligence, not with panic.

REFERENCES

1. Richard Feynman, *The Character of Physical Law* (Cambridge, MA: MIT Press, 1965), p. 148.
2. Richard P. Feynman, "*What Do You Care What Other People Think?*" (New York: W. W. Norton, 1988), pp. 135–138.
3. Richard A. Kerr, "Official Forecasts Pushed Out to a Year Ahead," *Science* 266 (1994):1940.
4. Pamela S. Zurer, "Nations Unlikely to Meet Even Short-Term Goals of Climate Treaty," *Chemical & Engineering News* 19 December 1994:28–29.
5. Ibid.
6. Royston M. Roberts, *Serentipity, Accidental Discoveries in Science* (New York: Wiley Science Editions, 1989).

Glossary of Terms

acid rain: precipitation that contains acid solutions or particles, formed as a result of sulfur dioxide emissions from industry.

adaptation: the process by which organisms adjust to a new or changed environment.

aerosol: very small particulate matter (liquid or solid) in the atmosphere.

albedo: fraction of the sun's radiation reflected back into space by a surface such as clouds, snow, aerosols, or water, providing a cooling effect to the atmosphere.

anthropogenic: caused or produced by human activity.

apocalyptic: referring to imminent destruction.

atmosphere: the blanket of gases that surrounds the earth.

biodiversity: the variety of plants and animals (biota) that inhabit the earth or particular ecosystems.

biomass (fuel): organic matter that can be converted to fuel and therefore be considered an energy resource. Biomass is carbon-based and will emit CO_2 gas when burned.

biota: the animals, plants, and microorganisms of a given area.

calibration: to set or check the mark, output signal, or measurements of an instrument to provide for consistency and accuracy.

carbon cycle: the very complicated process by which carbon is removed from the atmosphere and returned to it.

carbon dioxide (CO_2): the most commonly mentioned greenhouse gas, a compound made of carbon and oxygen.

CO_2 emissions: carbon dioxide gases put into the atmosphere; usually refers to emissions from combustion.

carbon sinks: the bodies or processes that remove carbon dioxide from the atmosphere and store it, such as plants, oceans, and lakes.

chaos: nonlinear, unpredictable behavior of systems; many natural systems exhibit chaotic behavior, including the atmosphere, oceans, and clouds.

chlorofluorocarbon (CFC): a classification of man-made chemicals, developed in the 1930s, that have been implicated in ozone depletion and global warming issues.

chromosphere: a thin shell of solar atmosphere, the area in which solar flares occur. The chromosphere cannot be seen from earth, but when observed from space, it exhibits tremendous activity.

clear-cutting: the practice of cutting all the trees in a given area at one time.

climate: the average of the weather conditions for a particular area, taken over a period of time.

climate-feedback variable: in climate models many variables, such as cloud cover or carbon dioxide levels, have either a warming effect (a positive feedback) or a cooling effect (a negative feedback) on predicted future climate.

climate model: a computer model that attempts to simulate the climate for the purpose of predicting future climate patterns.

climatology: the study of the earth's climate and climate conditions.

cloud: a visible mass in the atmosphere that can consist of all three water phases—ice, liquid, and vapor—as well as particulates and trace gases.

combustion: the process of burning. If a carbon-based fuel is burned, the combustion products are carbon dioxide gas and water vapor.

conservation: controlled utilization of resources in order to preserve or protect them.

convection: the movement of heat by circulation of heated parts of a liquid or gas; the vertical transport of liquids and gases of the oceans and the atmosphere.

convection currents: rising or sinking air or ocean currents that stir the atmosphere or ocean and transport heat from one area to another.

convection zone (of the sun): the area of the sun that is just outside the radiation zone, in which temperature, pressure, and density of the sun's radiation decrease enough to allow radiation to escape to the photosphere, where the sun's energy is emitted to space.

core (of the sun): the innermost zone of the sun, comprising about one-fourth of the sun's radius, where the nuclear fusion reactions producing the sun's energy take place. It is the hottest part of the sun and also the densest.

coupled model: a computer climate model that combines an atmospheric general circulation model (GCM) with a general oceanic GCM to simulate more realistically the oceanic–atmospheric interface.

cryosphere: the portion of the earth's surface that is made up of ice and snow—the ice sheets, glaciers, ice-covered bodies of water, and snow-covered land.

DDT: dichloro-diphenyl-trichloroethane, a toxic compound formerly used as an insecticide.

debunker: a person who seeks to expose something as being false or exaggerated.

deforestation: massive removal of forests from land. Deforestation results in the loss of important carbon sinks.

dendrochronology: the process of determining past climate by measuring the growth of tree rings.

developed country (nation): a highly industrialized country with a relatively high GNP and relatively low birth rate, such as the United States.

developing country (nation): a less industrialized country with a relatively low GNP and relatively high birth rate, such as India.

dimethyl sulfide (DMS): a sulfur-bearing gas produced by phytoplankton; it reacts in the atmosphere to produce sulfuric acid.

diurnal: occurring in daily cycles, such as day and night.

doomsayer: a person who dwells on imminent destruction, especially as caused by humanity.

Draconian: drastic, harsh, severe.

Earth Summit: a 1992 United Nations conference held in Rio de Janeiro, Brazil, to address environmental issues and pass a worldwide treaty dealing with greenhouse gases. Officially titled United Nations Conference on Environment and Development (UNCED).

eccentricity (of the earth's orbit): a cyclical change (about every 100,000 years) in the earth's orbit from elliptical to more circular that causes the earth to vary in its distance from the sun, causing a variation in the amount of sunlight impinging on the earth.

ecology: the study of the environment and how organisms interrelate with it.

ecosystem: a specific biological community and its environment.

eco-warrior: an extreme environmental activist who goes beyond peaceful, orderly means to achieve a goal of environmental protection.

El Chichón: volcano in Mexico that erupted in 1982, creating atmospheric effects lasting for at least four years.

El Niño: a natural, quasi-periodic increase in the sea-surface temperature of the ocean currents off the coast of South America, which can affect climate worldwide.

empirical research: verifiable research based on direct experiment or observation.

energy balance: the relationship between incoming solar radiation, outgoing terrestrial radiation, and all of the interactions that occur with it.

ENSO: the connection between the El Niño and Southern Oscillation phenomena.

environment: the total of all conditions, biological and nonbiological, that affect the life of an organism or population.

environmental activist: a person who advocates or works for protection of the environment, and sometimes for change in human activities that affect the environment.

evaporation: the process of converting a liquid into a gas.

extrapolation: a prediction of a future event or behavior based on past or present information.

flux: a continuous change, movement, or flow.

fossil fuel: a naturally occurring mixture of compounds containing carbon, hydrogen, and other elements that can be burned for fuel, such as coal, petroleum, and natural gas.

Fraunhofer lines: hundreds of dark lines in the visible spectrum of the sun that coincide with the atomic emission lines, or "fingerprints," of chemical elements. They were first catalogued in 1802 by Munich optician Joseph Fraunhofer, by means of a spectroscope.

general circulation model (GCM): a computer model that attempts to describe the circulation of the atmosphere (or the ocean) in terms of all climatic parameters, such as wind speed and direction.

geoengineering: technological methods of changing the radiation balance of the atmosphere, or otherwise altering the climate.

geothermal energy: energy that is produced by heat stored in rocks beneath the earth's surface. The heat may be obtained either from the rocks or from water heated by the rocks.

glacier: an extended mass of ice and snow that forms on land and survives for a long period of time. During the ice ages, glaciers covered much of North America.

global temperature: the average temperature of the earth's atmosphere at the surface of the earth at any given time.

global warming: a perceived increase in global temperature as a result of the release of greenhouse gases produced by human activities.

greenhouse effect: a warming effect on the earth's atmosphere that results from the capture of radiation by molecules of water vapor, carbon dioxide, and other gases. It is believed that without the greenhouse effect, the earth would be a cold and frozen wasteland.

greenhouse gas: any gas that exhibits the greenhouse effect; the most important of these gases include water vapor, carbon dioxide, and methane.

GWP (global warming potential): a measure of the importance of various gases as to their effectiveness as greenhouse gases with respect to CO_2, which is assigned a GWP of one.

heat sink: any large mass that absorbs heat and changes temperature very slowly as compared to the atmosphere.

hydrocarbon: a compound containing only hydrogen and carbon. Hydrocarbons are found in all fossil fuels.

hydrological cycle (water cycle): the global relationships of all the water phases (solid, liquid, gas) and the processes by which one phase changes to another.

ice age: a period in the earth's history when the significant portions of the earth's surface were covered by glaciers.

ice core measurement: a technique for measuring atmospheric gases of earlier centuries by cutting a core from the ice pack of a glacier and releasing the trapped gases into instruments that measure the quantities of the gases and relate them to the stratification of snow and ice.

infrared radiation (IR): low-energy radiation; heat can be felt from it, but it is not visible to the human eye. The earth radiates infrared radiation into space.

interglacial: warm period between ice ages.

laminar flow: smooth and orderly ocean current flow.

Little Ice Age: a cooling that took place in Europe and North America during the last 500 years.

Mauna Loa: the site of the observatory in Hawaii where the longest continuous measurements of atmospheric carbon dioxide have been made and recorded.

media: news media, i.e., newspapers, television, radio, and magazines.

meteorologist: a scientist who studies the atmosphere, weather, and climate.

methane (CH_4): one of the anthropogenic greenhouse gases, generated by domestic animals (especially through flatulence), rice paddies, biomass burning, landfills, coal mining, and gas drilling.

microclimate: the climate in a small area, such as a frost hollow or windswept ridge, that may be different from the general region.

Milankovitch theory: a theory developed in the 1930s by Milutin Milankovitch, a Serbian engineer, that explains how climate change is affected by variation in the solar energy reaching the earth.

model: a mathematical equation or series of equations that can be solved to predict the results of a real-life activity or process.

molecule: a combination of two or more atoms.

Montreal Protocol: a 1987 international treaty that called for the elimination of CFCs because of their implication in destruction of the ozone layer.

Mt. Pinatubo: volcano in the Philippines that erupted in 1991, creating atmospheric effects that lasted for several years.

mystery of the missing carbon: the imbalance between the amount of carbon dioxide emitted and the amount measured in the atmosphere.

nitrous oxide (N_2O): one of the greenhouse gases that has both natural sources (ocean evaporation and forest transpiration) and anthropogenic sources (soils on cultivated farmlands).

nuclear energy: the energy that is produced by a nuclear reaction.

nuclear fusion reaction: the reaction that occurs when two hydrogen atoms are fused together at extremely high temperature and pressure to form a helium atom, releasing energy. Nuclear fusion is the source of the sun's energy.

orbital variation: cyclical changes in the eccentricity of the earth's orbit, the precession of the earth's axis, and the wobble in the earth's axis that causes the earth to change its distance from the sun or it's angle to the sun, causing a variation in the amount of sunlight striking the earth.

ozone (O_3): a very reactive form of oxygen that attacks most chemicals in the atmosphere. It is a very minor greenhouse gas, but plays an important role in the stratosphere, where it absorbs high-energy ultraviolet radiation before it strikes the earth.

ozone hole: a decreasing of the ozone layer in the stratosphere over Antarctica, thought to be caused by CFCs.

part per million (ppm): a unit of concentration measure for indicating trace levels; e.g., 350 ppm of CO_2 means 350 molecules of CO_2 for every million molecules of total atmospheric gases.

particulate: a very small particle of matter, liquid or solid, in the atmosphere. Also called an aerosol.

passive microwave radiometry: a technique used to measure atmospheric temperature that employs satellite sensors (Microwave Sounding Units, or MSUs) to sense microwave emissions from the earth and its atmosphere.

permafrost: permanently frozen soil in the higher latitudes of the Northern Hemisphere that regulates the amount of water evaporated into the atmosphere in these areas.

perturbation: change; for example, a change in carbon dioxide levels is thought to cause a perturbation in the earth's climate.

photosphere (of the sun): the visible outer portion of the sun from which radiation escapes; this is where sunspots occur and where the sun's temperature is measured.

photosynthesis: the process by which plants utilize sunlight to create nutrients from carbon dioxide and water, forming oxygen as a by-product.

phytoplankton: microscopic plant organisms in aquatic ecosystems.

polar ice caps: the large regions of ice and snow that cover the areas surrounding the North and South Poles.

precession: variation in the angle of the earth's axis, which occurs with a 20,000-year cycle, causing a change in the angle at which the sun's radiation strikes the earth.

radiant energy: a form of energy (electromagnetic radiation); the energy received by the earth is in the form of radiant energy, including UV, visible, and IR radiation.

radiation: the term used to describe the electromagnetic radiation emitted from hot bodies, such as the sun.

radiation balance: the difference between the amount of radiation (energy) absorbed by the earth's surface and that reemitted into space. If the earth absorbs more radiation than it emits, the atmosphere will warm, and vice versa.

radiation zone (of the sun): the zone just outside the core of the sun. Radiation created in nuclear fusion reactions slowly works its way from the core to the convection zone, taking millions of years to do so because of the many interactions with charged particles along the way. Its density and temperature slowly decrease as the radiation travels from the core to the convection zone.

relative humidity: a measure of the amount of water vapor in the air relative to the maximum amount the air can hold at a given temperature.

salinity: amount of dissolved salt; saltiness. A term often used to describe different types of water in the ocean.

scenario: an imagined sequence of events; a predicted plan or possibility.

solar constant: the amount of energy from the sun impinging on 1 square centimeter of the earth's atmosphere, currently considered to be 2 calories/cm^2/minute. It is not really a constant.

solar cycle: periodic change (11.1-year cycle) in the number of sunspots.

solar spectrum: the intensity of solar radiation as a function of its wavelength or energy.

solar variation: variation in the radiation (energy) output of the sun caused by the solar cycles, a factor not included in most computer climate models.

Southern Oscillation: a periodic change in atmospheric pressure that occurs when there is a major warming of ocean surface temperature near South America, as in the El Niño phenomenon.

spectroscope: an instrument used to look at very small portions of the energy spectrum. A spectroscope was used by Fraunhofer to observe the dark lines in the sun's visible spectrum.

stratosphere: the layer of the earth's atmosphere between the troposphere (closest to the earth) and the mesosphere that is about 12–45 kilometers in altitude (40–150,000 feet).

sulfur dioxide (SO_2): a gas emitted by many natural and anthropogenic sources. It converts to sulfuric acid aerosol and is implicated in acid rain and in aerosol cooling in the stratosphere. Volcanic activity is thought to be responsible for the levels of sulfur dioxide in the stratosphere.

sunspots: large dark spots that appear on the surface of the sun in cycles and then disappear. They are associated with the energy output of the sun and therefore affect the earth's temperature.

supercomputer: the term given to the largest, fastest, most powerful computers. These computers are required to run computer climate models because of the great computational requirements that these models have.

tilt (of the earth's axis): the position of the earth's axis relative to the sun; the tilt angle changes slowly with a cycle of about 40,000 years, causing a variation in the amount of the sun's energy that hits the earth.

trace gas: any of the less common gases found in the earth's atmosphere, including carbon dioxide, methane, sulfur dioxide, nitrous oxide, or ozone.

tree ring measurement (dendrochronology): the process of determining past climate from the growth patterns of trees.

troposphere: the lowest layer of the atmosphere, extending up to an altitude of about 12 kilometers (40,000 feet).

turbulent flow: ocean current flow that is nonlinear, chaotic, and characterized by a great variety of speeds and eddies.

ultraviolet radiation (UV): the most energetic portion of the sun's radiant energy spectrum, which can cause damage to living tissue; it is the portion of the spectrum that, coupled with visible radiation, interacts with plants to provide energy for photosynthesis.

urban heat sink (or island): areas of human population concentration where vegetation has been replaced by asphalt and concrete, which absorb radiant energy and gradually increase the temperature of the area.

validation: a process used in computer modeling whereby one runs variables and compares them to past or present reality. If the model cannot predict the past or present reality with accuracy, it is not valid.

variable: a quantity or function that may assume any given value or values. Used in a computer model, a variable might be wind speed or direction, or level of carbon dioxide.

visible radiation: the portion of the solar spectrum that is visible to the human eye.

water cycle (hydrological cycle): the global relationships with respect to all water phases (solid, liquid, gas) and the processes by which one phase changes to another.

water vapor: the gaseous form of water, which varies significantly on a regional basis and with altitude.

weather: the state of the atmosphere, including temperature, moisture, atmospheric pressure, and wind.

wobble: small variation in the angle of the earth's axis, which changes from 21.5° to 24.25° with a cycle of about 40,000 years.

zooplankton: microscopic animal organisms in aquatic ecosystems.

Index

Riley